W9-AJM-090

-MASONRY-
PROJECTS &
TECHNIQUES

MASONRY PROJECTS & TECHNIQUES

Richard T. Kreh, Sr.

Sterling Publishing Co., Inc.   New York

Library of Congress Cataloging-in-Publication Data
Kreh, R. T.
    [Masonry projects and techniques]
    Masonry projects & techniques / by Richard T. Kreh, Sr. ; drawings
by Eugene Thompson . . . [et al.].
        p.   cm. — (Popular science)
    Reprint. Originally published: Masonry projects and techniques.
New York, NY : Popular Science Books ; Emmaus PA : Distributed by
Rodale Press, c1985.
    Includes index.
    ISBN 0-8069-6910-5 (pbk.)
    1. Masonry—Amateurs' manuals.   I. Title.   II. Title: Masonry
projects and techniques.   III. Series: Popular science (Sterling
Publishing Company)
TH5313.K725   1988   88-18130
693′.1—dc19   CIP

1 3 5 7 9 10 8 6 4 2

First published in paperback in 1988 by Sterling Publishing Co., Inc.
Two Park Avenue, New York, N.Y. 10016
Originally published in hardcover by Grolier Book Clubs, Inc.,
under the title "Masonry Projects and Techniques" copyright © 1985 by Richard T. Kreh, Sr.
Distributed in Canada by Oak Tree Press Ltd.
℅ Canadian Manda Group, P.O. Box 920, Station U
Toronto, Ontario, Canada M8Z 5P9
Distributed in Great Britain and Europe by Cassell PLC
Artillery House, Artillery Row, London, SW1P 1RT, England
Distributed in Australia by Capricorn Ltd.
P.O. Box 665, Lane Cove, NSW 2066
Manufactured in the United States of America
All rights reserved

Sterling ISBN 0-8069-6910-5 Paper

CONTENTS

Part II

MASONRY TECHNIQUES

WEIGHTS AND MEASURES

UNIT	ABBREVIATION	EQUIVALENTS IN OTHER UNITS OF SAME SYSTEM	METRIC EQUIVALENT
Weight			
Avoirdupois			
ton			
short ton		20 short hundredweight, 2000 pounds	0.907 metric tons
long ton		20 long hundredweight, 2240 pounds	1.016 metric tons
hundredweight	cwt		
short hundredweight		100 pounds, 0.05 short tons	45.359 kilograms
long hundredweight		112 pounds, 0.05 long tons	50.802 kilograms
pound	lb *or* lb av *also* #	16 ounces, 7000 grains	0.453 kilograms
ounce	oz *or* oz av	16 drams, 437.5 grains	28.349 grams
dram	dr *or* dr av	27.343 grains, 0.0625 ounces	1.771 grams
grain	gr	0.036 drams, 0.002285 ounces	0.0648 grams
Troy			
pound	lb t	12 ounces, 240 pennyweight, 5760 grains	0.373 kilograms
ounce	oz t	20 pennyweight, 480 grains	31.103 grams
pennyweight	dwt *also* pwt	24 grains, 0.05 ounces	1.555 grams
grain	gr	0.042 pennyweight, 0.002083 ounces	0.0648 grams
Apothecaries'			
pound	lb ap	12 ounces, 5760 grains	0.373 kilograms
ounce	oz ap	8 drams, 480 grains	31.103 grams
dram	dr ap	3 scruples, 60 grains	3.887 grams
scruple	s ap	20 grains, 0.333 drams	1.295 grams
grain	gr	0.05 scruples, 0.002083 ounces, 0.0166 drams	0.0648 grams
Capacity			
U.S. Liquid Measure			
gallon	gal	4 quarts (2.31 cubic inches)	3.785 litres
quart	qt	2 pints (57.75 cubic inches)	0.946 litres
pint	pt	4 gills (28.875 cubic inches)	0.473 litres
gill	gi	4 fluidounces (7.218 cubic inches)	118.291 millilitres
fluidounce	fl oz	8 fluidrams (1.804 cubic inches)	29.573 millilitres
fluidram	fl dr	60 minims (0.225 cubic inches)	3.696 millilitres
minim	min	1/60 fluidram (0.003759 cubic inches)	0.061610 millilitres
U.S. Dry Measure			
bushel	bu	4 pecks (2150.42 cubic inches)	35.238 litres
peck	pk	8 quarts (537.605 cubic inches)	8.809 litres
quart	qt	2 pints (67.200 cubic inches)	1.101 litres
pint	pt	½ quart (33.600 cubic inches)	0.550 litres
British Imperial Liquid and Dry Measure			
bushel	bu	4 pecks (2219.36 cubic inches)	0.036 cubic metres
peck	pk	2 gallons (554.84 cubic inches)	0.009 cubic metres
gallon	gal	4 quarts (277.420 cubic inches)	4.545 litres
quart	qt	2 pints (69.355 cubic inches)	1.136 litres
pint	pt	4 gills (34.678 cubic inches)	568.26 cubic centimetres
gill	gi	5 fluidounces (8.669 cubic inches)	142.066 cubic centimetres
fluidounce	fl oz	8 fluidrams (1.7339 cubic inches)	28.416 cubic centimetres
fluidram	fl dr	60 minims (0.216734 cubic inches)	3.5516 cubic centimetres
minim	min	1/60 fluidram (0.003612 cubic inches)	0.059194 cubic centimetres
Length			
mile	mi	5280 feet, 320 rods, 1760 yards	1.609 kilometres
rod	rd	5.50 yards, 16.5 feet	5.029 metres
yard	yd	3 feet, 36 inches	0.914 metres
foot	ft *or* '	12 inches, 0.333 yards	30.480 centimetres
inch	in *or* "	0.083 feet, 0.027 yards	2.540 centimetres
Area			
square mile	sq mi *or* m²	640 acres, 102,400 square rods	2.590 square kilometres
acre		4840 square yards, 43,560 square feet	0.405 hectares, 4047 square metres
square rod	sq rd *or* rd²	30.25 square yards, 0.006 acres	25.293 square metres
square yard	sq yd *or* yd²	1296 square inches, 9 square feet	0.836 square metres
square foot	sq ft *or* ft²	144 square inches, 0.111 square yards	0.093 square metres
square inch	sq in *or* in²	0.007 square feet, 0.00077 square yards	6.451 square centimetres
Volume			
cubic yard	cu yd *or* yd³	27 cubic feet, 46,656 cubic inches	0.765 cubic metres
cubic foot	cu ft *or* ft³	1728 cubic inches, 0.0370 cubic yards	0.028 cubic metres
cubic inch	cu in *or* in³	0.00058 cubic feet, 0.000021 cubic yards	16.387 cubic centimetres

Part I
MASONRY PROJECTS

Projects of masonry—concrete, brick, stone, and block—have many unique advantages. No other group of building materials offers masonry's combination of:

- *Permanency.* No material is as highly resistant to natural deterioration by water, fire, wind, or sun as masonry. Remember that most surviving structures from ancient civilizations are of masonry.
- *Beauty.* Masonry has a natural look in any setting. Actually, the materials themselves are of nature. Earth, air, water, and fire combine to make concrete, blocks, bricks, and even stone.
- *Low Maintenance.* When compared with other building materials, masonry products are almost maintenance-free.
- *Workability.* Masonry construction techniques, as you will see after reading Part II, are relatively simple and easy to learn. But, before going into masonry building techniques, let us look at various ways of using concrete, brick, block, and stone.

STONE PIERS, RUBBLESTONE WALL, AND WOOD-RAIL STEPS

This project is a practical solution to retain a terraced area in the yard or grounds and provide a set of steps to reach the higher level. The combination of stone and wood-rail steps creates a rustic appearance and blends in with the natural surroundings. The walls and steps in this project were the design of a homeowner handyman who wanted to create the effect of an oriental garden; note the stone lantern and statue on top of the two piers at the steps.

First drive several wood stakes where the walls will be built and attach a level line marking the wall lines. Level the line by attaching a line level to it or placing a regular level on top of a straight 2×4 held under the line. This will probably require the help of one other person.

Excavate the earth to the proper depth (24″ depth is shown on the plan; the correct depth for you depends on the frost line in your area) under the two piers and 16″ under the dry stone walls. It is a good idea to check with the building inspector in your area for frost depth. The earth for the dry stone walls is removed only for stability of the wall,

as it does not require any footing. Larger flat stones should be laid to serve as a base under the dry stone wall.

Mix the concrete for the footing in a ratio of 1 shovel portland cement to 2 parts sand to 4 parts crushed stone or gravel with enough water to make it workable. Throw a couple of pieces of old steel pipe or other metal in the hole to give the concrete extra strength. Fence wire also works well.

A single bag of portland cement should be more than enough to mix the footing, as you will have a supply of sand and crushed stone on the job anyway to build the stone piers.

Stone

Rubblestone (natural random stone of assorted sizes as found in the fields or on rock piles) will be used to build the walls and the piers. See the chapter on techniques of stone masonry for estimating tips on stonework. Try, however, to pick stone that is assorted in height, width, and shape, as it will make a more interesting wall and decrease the amount of cutting required.

3

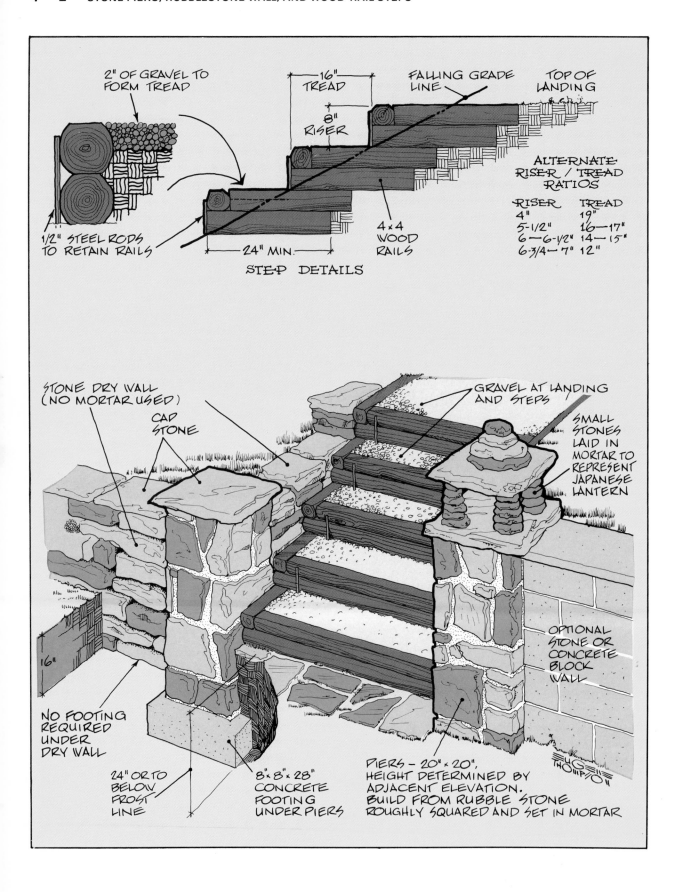

2" OF GRAVEL TO FORM TREAD

1/2" STEEL RODS TO RETAIN RAILS

16" TREAD

8" RISER

24" MIN.

STEP DETAILS

FALLING GRADE LINE

TOP OF LANDING

4 x 4 WOOD RAILS

ALTERNATE RISER / TREAD RATIOS

RISER	TREAD
4"	19"
5-1/2"	16—17"
6—6-1/2"	14—15"
6-3/4—7"	12"

STONE DRY WALL (NO MORTAR USED)

CAP STONE

GRAVEL AT LANDING AND STEPS

SMALL STONES LAID IN MORTAR TO REPRESENT JAPANESE LANTERN

16"

NO FOOTING REQUIRED UNDER DRY WALL

24" OR TO BELOW FROST LINE

8" x 8" x 28" CONCRETE FOOTING UNDER PIERS

PIERS – 20" x 20", HEIGHT DETERMINED BY ADJACENT ELEVATION. BUILD FROM RUBBLE STONE ROUGHLY SQUARED AND SET IN MORTAR

OPTIONAL STONE OR CONCRETE BLOCK WALL

UG THOMPSON

Mortar

A good mix for stonework is 1 part portland cement (Type 1) to 1 part hydrated lime to 6 parts sand. Add enough water to make the mortar to the stiffness desired. Stone mortar should always be stiffer than regular mortar for brickwork, because the hard stone does not absorb water and sets slowly. You will have to experiment until you find what consistency works best for you. Adding hydrated mason's lime to the mix is very important, because it makes the mortar more waterproof and more workable and prevents cracking of the joints.

Don't buy a lot of portland cement or lime to start with, because if it gets hard or wet the strength of the mix will be severely affected. Always put this material in a dry place off the ground. If you're laying any amount of stone at all, buy about 1 ton of sand, 4 bags of portland cement, and 4 bags of hydrated lime. Use these and then gauge how much more you will need based on that. The amount of mortar to build any stonework project will vary considerably depending on how many voids are in the wall and the thickness of the stone. For this reason, a definite amount of mortar for the project is not stated here. However, the chapter on techniques of stone masonry gives estimating standards for mortar for stonework.

Lay the stone piers up first, raking out the mortar joints with a short piece of broom handle to roughly ½" and then brushing well to close up all holes and present a finished appearance. Next, lay the dry rubblestone wall, chinking in areas between the larger stones with smaller ones. Try to lay flat flag-type stones on top of the wall to finish it and tie it together.

Lay the steps last. Lay the wood rails into position, tying them back into the earth bank (dig back into bank and tamp tightly in position with earth) to keep them from moving. Interlock them at the corners as shown in the drawing. As each rail step is laid into position, drive a length of ½" steel rod where needed along the front of the riser to hold the rail in place. Steel rods can be secured at building suppliers or junkyards and cut to specified lengths.

Next, fill in with gravel or crushed stone to form the tread of the steps. White gravel, available at most building supply or concrete companies, works well. The amount you will need, of course, depends on the length and depth of the treads. Gravel is a rather inexpensive material.

After all construction has been completed, fill in around the walls and steps with soil and plant a few shrubs to hold the earth. The result will be a beautiful set of steps with stonework that you can enjoy for many years.

HAND TOOLS

Bricklayer's trowel	2' square
Heavy-duty mason's hammer	Brush
	Marking crayon or pencil
Level	Pointing trowel (small
Ball of line	trowel)
Several masonry chisels	

EQUIPMENT

Mortar mixing box	Water hose
Wheelbarrow	Pick and digging iron
Shovel and mortar hoe	Mortarboard
Garden rake	

MATERIALS

(*Note:* Amount of materials in this project will depend on size of your wall and steps.)

Portland cement Type 1 (94-lb. bag)
Mason's hydrated lime (50-lb. bag)
Sand (sold by weight)
Short lengths ½" steel rods
Supply of rubble building stone
Crushed stone or gravel
Wood rails for steps (sometimes called garden timbers)

2
BRICK MAILBOX

A brick pier with a mailbox in it solves the problem of continuous replacement and can be a handsome addition to your home. However, be sure to check with the local postal authorities so as not to violate any restrictions and to find out the proper height for the mailbox. This is necessary to allow the letter carrier to reach out the window of his vehicle and insert the mail without getting out.

Select a location that does not infringe on the roadway. Excavate for the footing under the pier. This should be below the freeze line in your area. Call your county department for a building permit and find out what this depth is. Study the plans first and then excavate enough for placement of footings as shown on the plan.

After excavation, mix concrete in the following ratio: 1 part portland cement to 2 parts sand to 4 parts crushed stone or gravel, with enough water to make the mix workable. Place the concrete into the hole and tamp it level with a rake, leaving it rough on top. It is also a good idea to throw in some short lengths of metal pipe or wire for reinforcement.

Allow the concrete a day or so to set, then lay the concrete block on the footing as shown on the plan. Fill in the center with scrap block or old brick.

Lay out the brickwork as shown on the plan. Specific lengths and widths are not given on the plan, since there are minor variations in bricks. Use approximately ⅜" mortar head and bed joints between the bricks. Course the brick height to number 6 on the modular mason's rule.

Fill inside the pier with block as the work progresses, being careful not to push the brickwork out. When the correct height shown on the plans for the bottom of the mailbox is reached, insert a thin piece of cardboard around the box and set it into position in the pier. The cardboard is for expansion. Later you can pull this out and caulk around the outside edges of the box.

Set the brickwork in on each side as shown on the plan, 1½", to form the top of the pier. Continue laying brick until the last course is in place. Apply a mortar cove on an angle to prevent water from lying on the setback. Use solid brick for the top

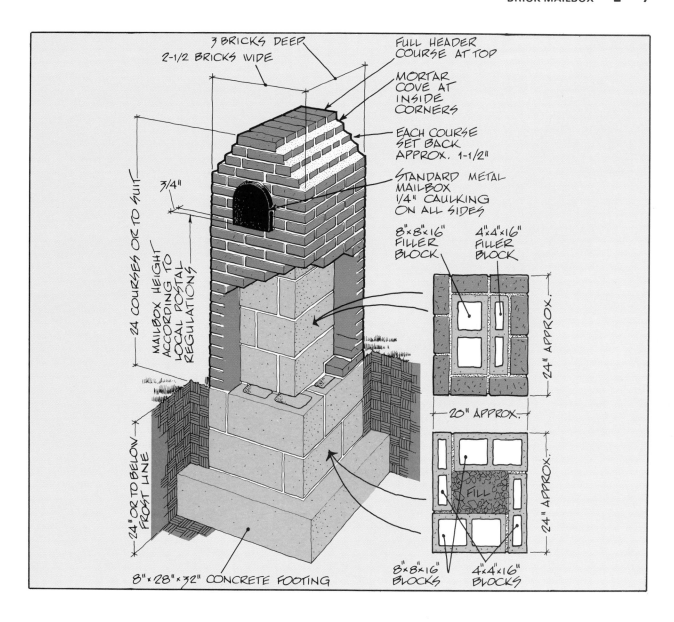

3 BRICKS DEEP
2-1/2 BRICKS WIDE
FULL HEADER COURSE AT TOP
MORTAR COVE AT INSIDE CORNERS
EACH COURSE SET BACK APPROX. 1-1/2"
STANDARD METAL MAILBOX 1/4" CAULKING ON ALL SIDES
3/4"
24 COURSES OR TO SUIT
MAILBOX HEIGHT ACCORDING TO LOCAL POSTAL REGULATIONS
8"×8"×16" FILLER BLOCK
4"×4"×16" FILLER BLOCK
24" APPROX.
20" APPROX.
FILL
24" APPROX.
24" OR TO BELOW FROST LINE
8"×28"×32" CONCRETE FOOTING
8"×8"×16" BLOCKS
4"×4"×16" BLOCKS

course and complete the pointing around the last course. Brush lightly after the mortar has dried enough to prevent smearing. Restrike, if necessary, to create a smooth, neat joint.

Clean the brickwork about a week later with a solution of 1 part muriatic acid (available at building supply dealers) to 10 parts water. Wet the pier well first, then scrub it with a brush and the acid solution. Follow with a good rinse to remove all of the dirt and mortar accumulation. Protect your hands with rubber gloves and wear a pair of safety goggles or glasses. If any of the acid solution gets on your hands or face, rinse it off immediately with water.

Last, the mailbox can be painted to add some color. Wait at least two weeks before planting any grass, to let the cleaning solution be absorbed into the ground. Now all that is necessary is to wait for the mail to be delivered.

HAND TOOLS

Bricklayer's trowel	Convex jointing tool
4' level	Pointing trowel
2' square	Brush
Brick chisel	Marking pencil
Brick hammer	Mason's Rule

EQUIPMENT

Mixing box	Bucket
Wheelbarrow	Hose
Mortar hoe and shovel	Pick and digging iron

MATERIALS

225 standard bricks
14 concrete blocks, 8"×8"×16"
3 bags masonry cement
¾ ton building sand
1 bag portland cement (for footing)
Approx. 200 lbs. crushed stone (for footing)
1 standard mailbox
1 tube caulking
Piece of cardboard to insert around mailbox in brickwork

BRICK BASKET PLANTER

This brick basket planter looks difficult to build but in truth is only a simple application of basic bricklaying techniques. It is a real eye-catcher when built on the front lawn and full of lovely summer flowers. It is a project that should not be rushed; the curved walls must be leveled and plumbed carefully to retain its shape. The end result will be well worth the effort.

Start by excavating for the footing. The simplest way to do this is to excavate the entire area, as it is only a little over 6′ in diameter. It will be easier to work if all of the earth is out of the way. Make sure footing is down to the freeze line in your area.

Mix and pour a footing for the wall of 1 part portland cement to 2 parts sand to 4 parts crushed stone. Determine the location of the footing by laying a garden hose in a circle approximately where the wall should be. Drive level stakes in the center of the footing area. Pour the concrete level with the stakes and then pull them out.

Since the rounded wall is only 4″ wide (one brick in width), I would start with brick right on the footing and lay out the bond so that it works with whole bricks all the way around the circle. Notice

on the plan that there is a double brick pier on each side of the center of the project. This will be the pier that supports the basket handle when that point is reached later. See photo for this.

Note: If you have some old chipped bricks, they will work fine under the grade line. However, make sure that your good bricks are laid out about two courses of bricks below the finished grade to assure a correct bond.

Continue building the project up to one course above finished grade line. Insert a piece of plastic tubing every third brick in the bottom of the head joint around the project to allow excess water to drain (see detail on plan). Tool the mortar head and bed joints with the convex striker.

Continue building up to the height where the piers start, form these, and lay bricks to the point where the angle irons are to be set in place (see detail on plan). Try to select bricks of the same length for the piers so that none jut out beyond the others.

Lay the angle irons in a back-to-back position as shown on end-view plan. Angle irons should rest on each pier a minimum of 4″ for good support. You

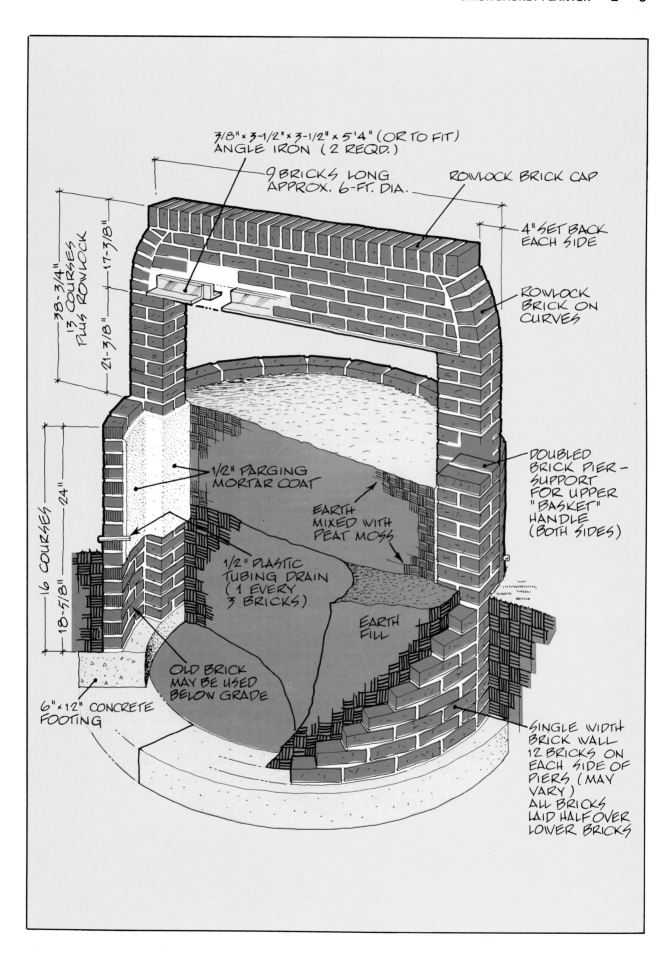

3/8" × 3-1/2" × 3-1/2" × 5'4" (OR TO FIT) ANGLE IRON (2 REQD.)

9 BRICKS LONG APPROX. 6-FT. DIA.

ROWLOCK BRICK CAP

4" SET BACK EACH SIDE

ROWLOCK BRICK ON CURVES

38-3/4" 13 COURSES PLUS ROWLOCK

17-3/8"

21-3/8"

24"

16 COURSES

18-5/8"

1/2" PARGING MORTAR COAT

EARTH MIXED WITH PEAT MOSS

1/2" PLASTIC TUBING DRAIN (1 EVERY 3 BRICKS)

EARTH FILL

DOUBLED BRICK PIER – SUPPORT FOR UPPER "BASKET" HANDLE (BOTH SIDES)

OLD BRICK MAY BE USED BELOW GRADE

6" × 12" CONCRETE FOOTING

SINGLE WIDTH BRICK WALL 12 BRICKS ON EACH SIDE OF PIERS (MAY VARY) ALL BRICKS LAID HALF OVER LOWER BRICKS

will have to measure the exact length and buy them or have them cut to this length for your project, as the length could vary depending on brick lengths. I would strongly recommend buying heavy steel angle iron of ⅜" thickness; pressed-steel irons will bend under the load. Also, apply a coat of rust-preventive paint on the irons before installing.

Make a pattern of cardboard or wood with the curvature as shown on the plans for the ends. This will be a 4" setback from the outside wall line (see detail on plan).

You can then determine where to cut the whole bricks to allow laying the rowlock outside course by holding the pattern against the end of the wall.

Lay the cut bricks in place as shown on the plan and then with the aid of the line fill in the wall that forms the handle.

Next, lay the curved rowlock bricks on the ends to fit the curve made by the pattern.

Lay the rowlock brick top to the line as shown on the plan. Tool all mortar joints on top flat with the slicker tool and brush when dry.

Parge the inside of the flower basket with mortar up to the height where the earth will be filled to help make it waterproof (see plan).

Let the brickwork set at least three days and wash down with a solution of 1 part muriatic acid to 10 parts water. Be sure to flush the work with plenty of running water from a hose to get all of the chemicals off after the cleaning is done.

I would recommend waiting for three or four days before putting any earth inside the planting area to give the acid solution a chance to evaporate.

HAND TOOLS

Brick trowel	Brick chisel
Brick hammer	Convex jointer (rounded)
Mason's rule	Slicker jointer
4' level	Pointing trowel
2' level	Brush

EQUIPMENT

Mixing box	Mortarboard
Wheelbarrow	Pick and shovel
Shovel and hoe	Wire cutters
Bucket and hose	

MATERIALS

For footing

2 bags of portland cement
About ¼ ton sand
About 600 lbs. crushed stone

For Walls

500 standard bricks (should be solid with no holes)
5 bags masonry cement (includes enough to parge inside of wall as shown on plan)
½ ton building sand
1 masonry metal wall reinforcement, 10' length, for 8" wall
2 angle irons, 3½"×3½"×⅜", approx. 66" long (*Note:* This length will have to be determined by how long your brick works out across the opening at handle of basket as shown on plan.)
5' vinyl or plastic tubing, ½" diameter.

4

STONE POOL AND FOUNTAIN

There is nothing more relaxing than sitting next to a fountain of water on a warm evening and enjoying a nice cool drink. This project will turn your backyard into an enjoyable retreat. The fountain and pool in this project are very simple in design and require only very basic masonry skills. The addition of a circulating pump and a filter keeps the water clean and fresh. The homeowner who built this project created a very warm effect by illuminating the water and stonework with colored lights. If you want to do this, remember to build wiring into the masonry walls, keeping in mind that it will have to be replaced sometime. The cost is very reasonable and will show off your creativity and skill to your neighbors and friends.

Stake off the area and excavate the pool area as shown on the plan. The concrete floor and walls of the pool are reinforced with steel rods for extra strength to withstand the dampness. Because of the amount of concrete needed, it would be a good idea

to rent or borrow a small utility mixer to save your back.

Notice on the plan that the concrete walls are poured up to a height of 16" above the floor of the pool. This is done to make a more waterproof bond between the floor and the wall. My recommendation is to pour the floor first and then immediately set the forms for the wall on the floor and continue pouring the walls. This will unite all of the concrete in one solid mass. Be sure to have short stub reinforcing rods ready to insert into the floor and extend up into the walls. Some pieces of ⅝" or ¾" plywood will serve as a good economical form. This is shown on the plan.

Mix the concrete a little richer for this job than normal: a ratio of 1 part portland cement to 2 parts sand to 3 parts crushed stone. Let the concrete cure for a couple of days to prevent chipping, then remove the forms. Tapping on the forms lightly with a hammer when the concrete is being poured will

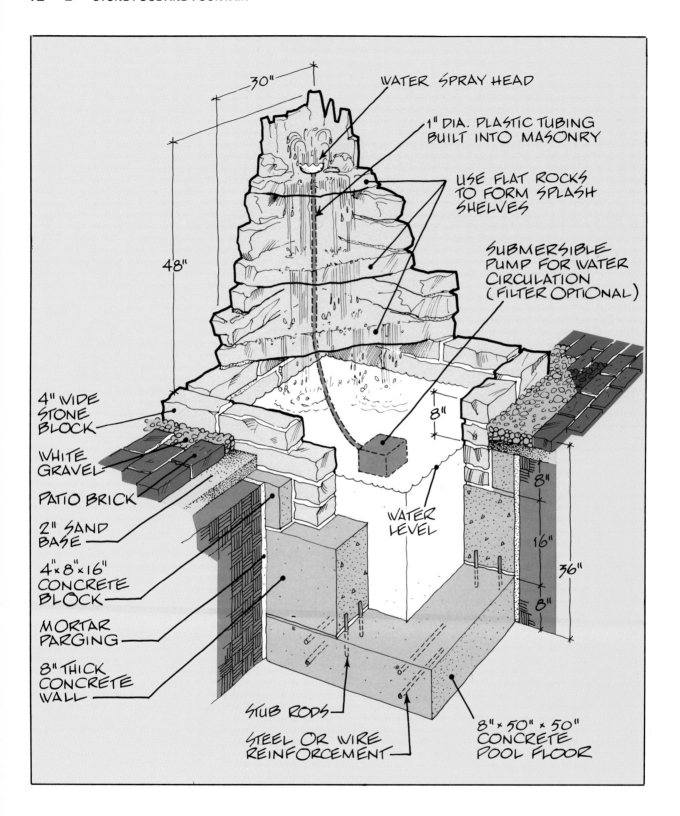

30"

WATER SPRAY HEAD

1" DIA. PLASTIC TUBING
BUILT INTO MASONRY

USE FLAT ROCKS
TO FORM SPLASH
SHELVES

SUBMERSIBLE
PUMP FOR WATER
CIRCULATION
(FILTER OPTIONAL)

48"

4" WIDE
STONE
BLOCK

WHITE
GRAVEL

PATIO BRICK

2" SAND
BASE

4" x 8" x 16"
CONCRETE
BLOCK

MORTAR
PARGING

8" THICK
CONCRETE
WALL

8"

WATER
LEVEL

8"

16"

36"

8"

STUB RODS

STEEL OR WIRE
REINFORCEMENT

8" x 50" x 50"
CONCRETE
POOL FLOOR

help to prevent any honeycombing on the surface. A coating of oil on the form boards is also a good idea, especially if you wish to reuse the forms.

Mix the mortar for the stone and block work: 1 part portland cement to 1 part hydrated mason's lime to 6 parts sand. This is known as Type N mortar and has excellent bonding qualities, which are desired for stonework.

Lay the stone around the perimeter of the pool as shown on the plan, making sure that all of the mortar joints are well filled. Smooth and tool all mortar joints with the slicker tool (flat tool) before they are dry. Brush lightly after they are dry to avoid smearing.

To form the angled back side of the pool where the waterfall and fountain will be, lay the stone across on an angle of approximately 45° as shown on the plan. Several stones laid back to back will be required to accomplish this.

Project these shelf stones to serve as a landing for the waterfall, as shown on the plan and in the photograph. Build a length of 1″ plastic tubing into the back corner as the stonework progresses for the water from the circulating pump to pass through.

If you want any electrical wiring built into the walls for lighting or for a pump or filter, it would be best to talk to an electrician or your local garden shop. The pumps and filters are usually available from Sears, Roebuck or Montgomery Ward at a reasonable price.

The top is cleverly built by laying some rather narrow stones in a vertical position, making sure that they are mortared in fully to withstand the excessive moisture. The continuous circulation of the water in the pool and through a filter keeps it fresh and clean. Finish this project by applying a blue or light-green waterproofing cement paint to add some color.

When the floodlights are on, the water will take on a beautiful color and the cement paint will help to prevent excessive water loss through the walls and floor. Cement paint is available at almost all building supply stores in a host of colors.

If you live in an area of the country where the temperature drops below freezing, be sure to remove the pump and filter and to drain the pool in the fall of the year. If water is left in the pool over the winter, throw a length of firewood in the water to prevent the water from freezing and cracking the walls.

The homeowner who designed and built this pool and fountain is a college professor of music. This project shows what a creative mind and a mastery of the basic skills of masonry can accomplish, and at a negligible cost.

HAND TOOLS

Brick trowel	Brush
Pointing trowel	Slicker jointing tool
Brick hammer	Square
4′ level	Marking crayon or heavy-
2′ level	duty carpenter's
Wide-blade chisel (called	pencil
brick set)	Concrete finishing or
Line blocks	plastering trowel

EQUIPMENT

Mixing box	Mortarboard
Wheelbarrow	Bucket and hose
Shovel and hoe	

MATERIALS

For Footing and Concrete Wall in Bottom Area of Pool

5 bags portland cement

1,500 lbs. sand

1,750 lbs. crushed stone (*Note:* These amounts will make 1 cubic yard of concrete, enough to pour floor and walls up to the 16″ height shown on plan. You should secure a utility mixer, as this would be too much to mix with a hoe. An optional choice would be to order 1 cubic yard of mixed concrete, but this would be expensive.)

4 pieces ¾″ steel reinforcement rods 4′6″ long or a section of concrete wire measuring 4′6″×4′6″ to be placed in the concrete when pouring. (It is very important that this concrete slab be reinforced, because of the damp conditions.)

For Stone Walls

1,500 lbs. ashlar stone (1 square foot of stone weighs about 50 lbs.)

12 concrete blocks, 4″×8″×16″, for backing up stone wall below grade.

2 bags portland cement

2 bags hydrated mason's lime

½ ton sand

5
BRICK CULVERT WALL

Begin by staking out and excavating where the footing will go. Be sure it is deeper than your local frost depth. Mix the concrete in a proportion of 1 part portland cement to 2 parts sand to 4 parts crushed stone.

Start laying brick directly off the footing. Lay a rowlock course (brick on edge) around the circular metal pipe, spacing the bricks out with the mason's rule, using the scale side. Number 6 on the modular scale should work. Squeeze back or open slightly to work even brick around the circle.

Attach a line stretched from one end to the other to make sure the wall lines up as it is built. The brickwork should first be dry-bonded to work out with whole bricks if possible. This is very important, as whole bricks should be used when you cross the top of the circular pipe. See plan for this.

Build in the metal joint reinforcements as shown on the plan to bond the wall together. This should occur every 16″ in height. Tool the mortar joints with a convex (round) jointing tool as soon as they

start to dry. Brush to compete the striking procedure.

The last course of brick is a double rowlock and should be half-bonded to lock the bricks into place better. See detail on plan for this. Tool the top mortar joints flat with a slicker tool or pointing trowel.

The brickwork can be cleaned after three days with a solution of 1 part muriatic acid to 10 parts water. Rinse with plenty of water with a hose to remove any chemicals or dirt remaining.

Finish the job by filling in with earth and raking the earth. A red or yellow reflector on a metal rod placed near the end of the wall to face turning traffic will help keep anyone from running into the wall. These are available at almost any hardware or general department store that stocks outdoor items.

The brick culvert wall will not only solve your water problem at the end of the driveway but add some charm to your property.

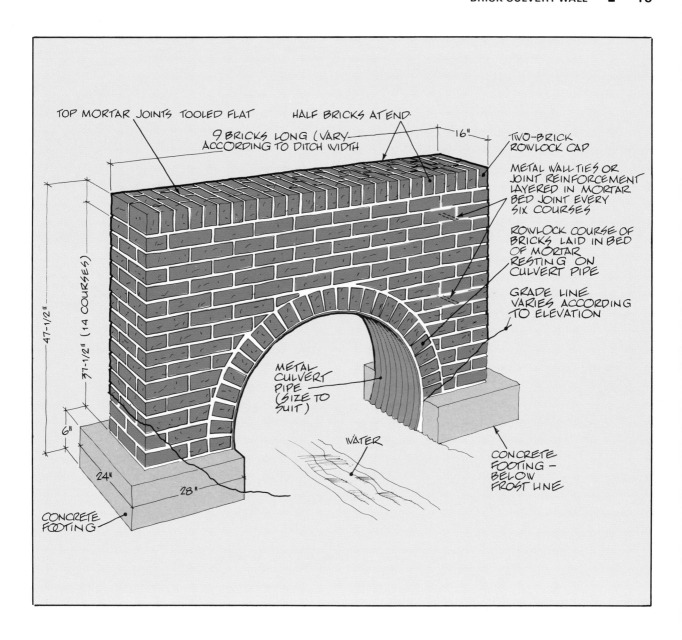

TOP MORTAR JOINTS TOOLED FLAT

HALF BRICKS AT END.

9 BRICKS LONG (VARY ACCORDING TO DITCH WIDTH

16"

TWO-BRICK ROWLOCK CAP

METAL WALL TIES OR JOINT REINFORCEMENT LAYERED IN MORTAR BED JOINT EVERY SIX COURSES

ROWLOCK COURSE OF BRICKS LAID IN BED OF MORTAR RESTING ON CULVERT PIPE

GRADE LINE VARIES ACCORDING TO ELEVATION

47-1/2"

37-1/2" (14 COURSES)

METAL CULVERT PIPE (SIZE TO SUIT)

WATER

6"

24"

28"

CONCRETE FOOTING

CONCRETE FOOTING – BELOW FROST LINE

HAND TOOLS

Brick trowel
Brick hammer
Mason's rule
4' level
Square
Brick chisel

Ball of line
2 line blocks
Convex jointing tool
(round-edge)
Brush
Pointing trowel

EQUIPMENT

Mixing box
Shovel and mortar hoe
Wheelbarrow
Bucket and hose

Mortarboard
Pick and digging iron
Rake
Wire cutters

MATERIALS

For Footing

1 bag portland cement
1 wheelbarrow sand
1 wheelbarrow crushed stone

For Wall

Approx. 300 standard bricks
3 bags regular masonry cement
½ ton sand
12' metal joint reinforcement, 12" lengths (to be laid in the bed joints as shown on the plan.) (*Note:* Steel culvert pipe is available from your local building supplier or junkyard in different sizes. Check to determine the size you will need. This will depend on the amount of water to be carried and the depth of your driveway edge.)

6
BRICK HERRINGBONE PANEL WALL

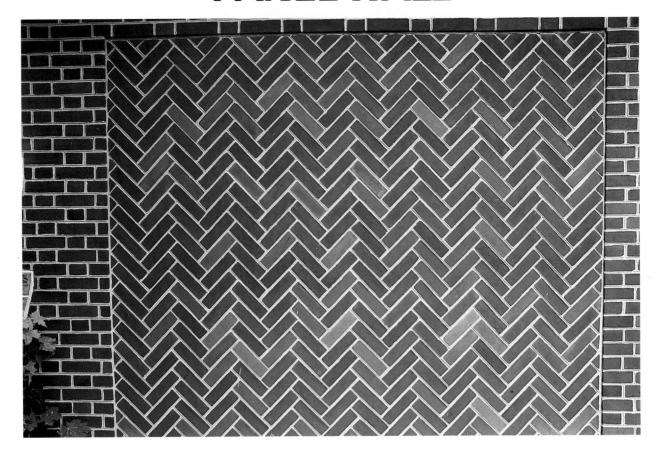

Here is a masonry project that will really test your mastery of the bricklaying skills. The key is to take your time and make sure that each and every brick is laid as shown on the plan and in the photograph. Although the project may look complicated and difficult, in truth it is not!

The herringbone pattern is started out as shown on the detail and merely is repeated throughout the project with care taken not to let the bond shift or get out of line.

Begin this project by marking off and excavating the footing area. Mix concrete for footing: 1 part portland cement to 2 parts sand to 4 parts crushed stone. Pour it in place. Let the footing harden for at least one day before laying any blockwork.

Begin your masonry work by laying the concrete block foundation up to the finished grade line. Use masonry cement mortar in a ratio of 1 part masonry cement to 3 parts sand.

The herringbone panel shown in the photograph was built into a Flemish bond brick wall and front-

ed on the patio area. You may design your adjoining wall however you wish. Remember there should be at least 1½ bricks on both sides of the panel to form a suitable border.

After the border walls are built (a cross-sectional plan of a typical wall is shown), the back of the panel should be built before attempting to lay any of the brick herringbone pattern. This is laid up of 4" concrete block, and wall ties are imbedded in the mortar bed joints every 16" in height and approximately 32" apart horizontally. When the brick panel work is laid, the wall ties will be mortared into it to bond the wall together. Keep the block good and plumb, as there is little room for error. Build the blockwork up to the height shown on the plan or so that whole bricks can be used at the top. Check with the mason's rule and work this out before laying any of the bricks.

You are now ready to start laying out the herringbone pattern. Start by making the cuts shown on the detail for the border. All cuts are made on a

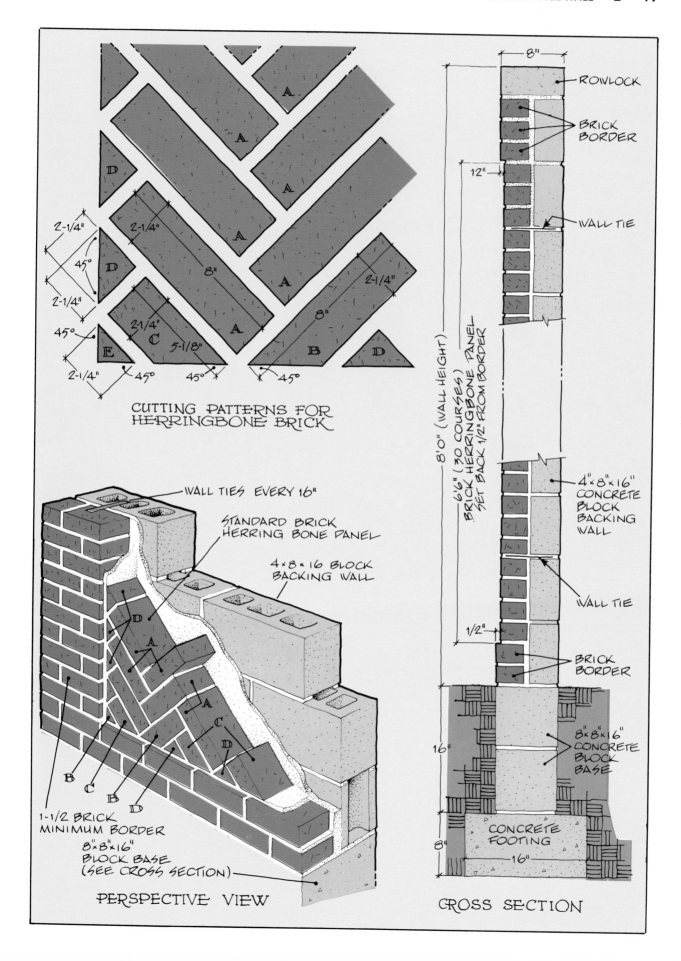

CUTTING PATTERNS FOR HERRINGBONE BRICK

WALL TIES EVERY 16"

STANDARD BRICK HERRING BONE PANEL

4×8×16 BLOCK BACKING WALL

1-1/2 BRICK MINIMUM BORDER

8×8×16" BLOCK BASE (SEE CROSS SECTION)

PERSPECTIVE VIEW

ROWLOCK

BRICK BORDER

WALL TIE

8'0" (WALL HEIGHT)

6'6" (30 COURSES) BRICK HERRINGBONE PANEL SET BACK 1/2" FROM BORDER

4"×8"×16" CONCRETE BLOCK BACKING WALL

WALL TIE

BRICK BORDER

8"×8"×16" CONCRETE BLOCK BASE

CONCRETE FOOTING

CROSS SECTION

45° angle. Mark the bricks with the combination square, which has a 45° angle setting on it. These are available at almost any store that stocks tools.

Make your cuts very accurately, using a broad-bladed brick chisel, commonly called a brick set. Be careful not to chip the face of the bricks, as it will be very noticeable. A good method to use to prevent breaking a lot of bricks is to lay them on a burlap bag of sand when cutting. This will cushion them and help prevent cracking.

Throughout the panel, you will notice the same cuts are made over and over to form the herringbone pattern. Also note that in the plans the panel is recessed ½" back from the face of the adjoining walls. Maintain this setback as the brick is laid.

Lay up each corner until you have a full brick point touching the vertical border (see photo). This will be the level point where the line should be attached for laying out the bond. This is critical in maintaining the pattern throughout the project.

Raise the line when the next set of full bricks is reached and continue to build the herringbone panel. Fill gently with mortar in back of the brickwork to cement the brick and block backing together.

As the height of the wall ties is reached, be sure to mortar them fully into the brickwork.

Check with the mason's rule to make sure that the brick works out even with the top of the panel as shown in the photograph

Tool the mortar joints as soon as they are thumbprint-hard. The project shown in the photograph has a grapevine joint for the finish.

Not all bricks are the same exact length, as you will soon discover. There is a slight difference between some bricks because of the burning and shrinkage in the kiln. Avoid using bricks markedly longer or shorter than average. This should be no big problem if you bought 500 bricks as mentioned in the materials listing. You may also have to vary the mortar head and bed joints slightly to effect a consistent bond pattern.

When the top of the panel is reached, it is not necessary to install an angle iron to support the bricks that cross over it. You should, however, set the brick courses back out the ½" to match the

original work under the panel. This is shown on the plans.

Complete the job by washing down with a solution of 1 part muriatic acid to 10 parts water after it has cured for at least three days.

This project will take more patience and time than many, but will be a most satisfying development of your skills and a conversation piece to anyone who sees it.

HAND TOOLS

Brick trowel	Grapevine jointer
Brick hammer	Combination square for
Mason's rule (modular)	making 45° cuts
4' level	Regular square
2' level	Ball of line
Bricklayer's pointing	Line blocks
trowel	Brush

EQUIPMENT

Mixing box	Bucket and hose
Wheelbarrow	Mortarboard
Shovel and hoe	Pick and digging iron

MATERIALS

(*Note:* The amount of concrete, concrete block, brick, and other materials will depend on how long your project will be. The amounts given here are only for the panel wall as shown on the drawing and in the photograph and do not include any of the adjoining walls, as they are not considered a part of this project. The wall as shown actually requires 425 bricks based on seven bricks per square foot of wall area. Because of the need to cut pieces around the border to effect the bond and the possibility of cracked or chipped bricks, I am recommending buying 500 bricks. This will also permit a selection of bricks for color, size, and variation.)

For Footing
2 bags portland cement
400 lbs. sand
500 lbs. crushed stone

For Walls
500 standard bricks
6 concrete blocks, 4"×8"×16", for backing up the brick panel
6 bags masonry cement mortar, standard Type N
¾ ton sand
50 corrugated veneer wall ties, approx. 7" long by 1" wide and 1/16" thick (available at building supply dealers)

7

BRICK ARCHED OPENING AND WALL

This project is a real challenge but will be worth all the time and effort you put into it. The brickwork and arched opening shown in the photograph were done by a handyman who is a college professor of music and not a working mason. It will give him many pleasant hours of enjoyment and satisfaction over the years.

Begin construction by excavating and pouring the footing. Mix the concrete to proportions of 1 part portland cement to 2 parts sand to 4 parts crushed stone.

After the footing has cured for at least a day, lay the two courses of block up to the finished grade line as shown on the plan. Even though an opening is shown, build the block wall through the opening for strength. This will also form a stronger support for the wall.

Dry-bond the brick from one end to the other to establish the bond. Form approximately ⅜" mortar head joints. Next, lay a brick on each end of the wall in mortar to the correct height (number 6 on modu-

lar scale rule), attach a line, and lay the first course through to the line. Leave out the opening.

As the brick wall is built, lay in lengths of 8" metal joint reinforcement in the bed joints every 16" in height. This would be every six courses of brick. Cut the reinforcement to fit, as it comes in 10' lengths. Be sure to mortar the reinforcement in fully for best results.

The bricks at the jambs (edges of the openings) are laid in a header position (the long way) and have to be kept especially level and plumb. Additional wall ties can be inserted here for extra strength. See front elevation of plan.

Lay the brick courses up to a height of 50⅝". This will be 19 courses. Set the wall back at that point as shown on the front elevation plan. Build three more courses of brick up to the turning point shown on the plan.

Cut out arch forms as shown on detail and set into position, bracing the center. It is important that the beginning edge of the arch curvature be exactly

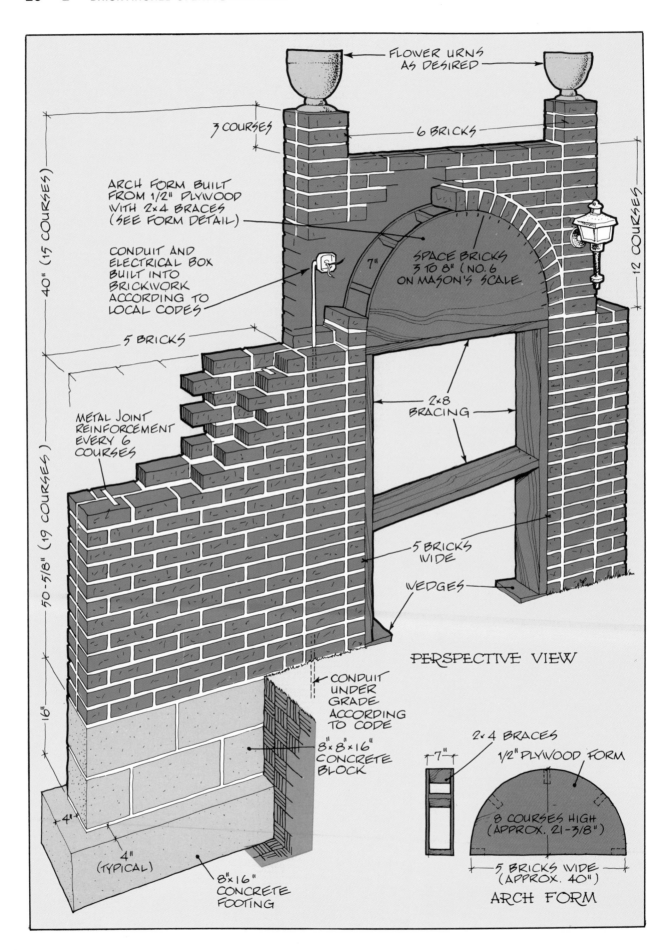

FLOWER URNS
AS DESIRED

3 COURSES

6 BRICKS

40" (15 COURSES)

12 COURSES

ARCH FORM BUILT
FROM 1/2" PLYWOOD
WITH 2×4 BRACES
(SEE FORM DETAIL)

CONDUIT AND
ELECTRICAL BOX
BUILT INTO
BRICKWORK
ACCORDING TO
LOCAL CODES

7"

SPACE BRICKS
3 TO 8" (NO. 6
ON MASON'S SCALE

5 BRICKS

2×8
BRACING

METAL JOINT
REINFORCEMENT
EVERY 6
COURSES

50-5/8" (19 COURSES)

5 BRICKS
WIDE

WEDGES

16"

PERSPECTIVE VIEW

CONDUIT
UNDER
GRADE
ACCORDING
TO CODE

8"×8"×16"
CONCRETE
BLOCK

2×4 BRACES

7"

1/2" PLYWOOD FORM

8 COURSES HIGH
(APPROX. 21-3/8")

4"

4"
(TYPICAL)

8"×16"
CONCRETE
FOOTING

5 BRICKS WIDE
(APPROX. 40")

ARCH FORM

level with the last course of brick at the turning point if it is to turn correctly. Insert wood wedges under the bottom of the 2×8 form braces to allow adjustment. This also helps when removing the forms later after the work has been completed and cured.

Around the top edge of the arch form, mark off individual brick spacing with the rule and the pencil. Use number 6 on the modular rule scale and adjust any difference by forming a slightly thicker or thinner mortar head joint. This is shown on detail of arch form in position on plan. You should lay the arch bricks from both bottom ends ot the center, inserting the last (key) brick with a double mortar joint so that the arch will not leak water through.

The brickwork on both sides of the arch should be laid as the arch is being built to tie everything together well. If all of the work is done correctly, the brick should course over the top of the arch with no cutting for height as shown on the plan. If not, work it out neatly so that is does not look objectionable.

Caution: Do not remove the arch form after the arch is in place until at least three days have passed, as the brickwork could drop and crack. Once all of the brickwork, including the arch, has cured, it will be stronger than any type of lintel that you could have installed.

Continue laying the brick above and across the arch as shown on the plan. The first course of brick laid over the head of the arch should have a length of metal joint reinforcement built into the mortar bed joints for extra strength. Tool the mortar joints as needed and brush lightly.

The last three courses form a post on each side of opening on which an urn or other object can be placed if so desired. The homeowner who built this project liked special effects.

If lights are desired as a part of the wall as shown in the photograph, be sure to build in metal electrical pipe conduit as the wall is constructed to protect the wire. In addition, build an electrical box into the brickwork to which you can attach the fixture. Ask an electrician how this should be done.

As previously stated, after three days have passed, remove the forms gently and point up the mortar joints on the bottom interior of the arch. A small chisel works well for trimming the joints, and a pointing trowel will finish the job. Tool the re-pointed joints with the same joint finish used on the face of the brickwork.

Let the wall cure for about a week, then clean it by washing it down with a solution of 1 part muriatic acid to 10 parts water. Wear safety eye protection and rubber gloves, and rinse the wall well before and after cleaning to remove all dirt and chemicals.

HAND TOOLS

Bricklayer's trowel	2' level
Brick hammer	Mason's modular rule
4' level	Convex jointer
Ball of line	Slicker jointer
2 line blocks	Brick set chisel
2' square	Brush

EQUIPMENT

Mixing box	Mortarboard
Wheelbarrow	Bucket and hose
Shovel and hoe	Pick and digging iron

(*Note:* If you build the arch form over the opening, you will need standard carpenter's tools such as a claw hammer, handsaw, small square, etc., and a supply of 8d and 6d common nails to fasten the form together. If you hire a carpenter for this, of course, you will not need these tools.)

MATERIALS

(*Note:* The arched opening shown in this project is a section of wall. The intersecting wall on the right, seen in the photo, is not shown on the plan as a part of the project. The intent here is to show how a typical section of wall with an arched opening is built. Materials listed are only for the section of wall with an arched opening. For reference on estimating brick, mortar, etc., see the chapter on techniques of brick masonry.)

For Footing
2 bags portland cement
400 lbs. sand
500 lbs. crushed stone

For Wall
14 concrete blocks, 8"×8"×16"
575 standard bricks
6 bags regular masonry cement
¾ ton sand
30 lineal feet 8" metal joint reinforcement

(*Note:* If you make your own arch form, you will also need a piece of ½" plywood approximately 16"×40". Two forms are cut, one for each side, in the curvature of the arch. Use short pieces of 2×4 or 1×3 to brace the arch form inside. See details of form construction on plan. Last, you will need two 2×8 boards about 7' long to support the arch form once in position. See detail on plan of how this is made and braced into position.)

8
CONCRETE BLOCK RETAINING WALL

oncrete block is fairly inexpensive when compared to forming and pouring concrete. If the blockwork is done neatly and the wall painted with a waterproof paint, when completed it presents a pleasing appearance. Most concrete block retaining walls that fail do so because they were not reinforced with steel and concrete. There is a tremendous amount of pressure exerted on a retaining wall from water and movements of the earth during freezing and thawing.

Before starting on this project, try to remove any future problems that may affect the stability and success of the wall. Number one, in my opinion, is to make sure that there are no downspouts from the rain gutter emptying behind where the wall will be built. If there are, extend them out away from the wall by connecting them to drain tile or drain pipe. Dig and place them under the finished grade line to make a nice-looking job. These pipes and tiles are available from your local building supply dealer.

Stake out the wall and attach a line. Excavate the wall area to the footing depth required by the freeze line in your area. Study the typical wall section on the plan and notice that this footing is con-

siderably wider than for a normal wall. This is the secret of building a strong retaining wall. Tests conducted by the Brick Institute of America prove that a wide footing for retaining walls greatly helps prevent the walls from moving. Because the footing is wider on the inside of the wall, it exerts a cantilever effect. The earth on the inside of the wall exerts an enormous amount of pressure against the footing, acting as a wedge to prevent the wall and footing from shifting forward. Although it may seem a little costly to install this footing as shown, it will pay off in the long run.

Start by mixing and pouring the footing, using a mix of 1 part portland cement to 2 parts sand to 4 parts crushed stone. Insert lengths of steel rod in the approximate center of the proposed wall line, letting them project up vertically about 7". Space them about 4' apart. These will act as stub tie rods for the long rods that will be inserted later in the hollow cells of the concrete blocks after the wall is built up to its height. See detail on plan.

Lay the first course of block in mortar, making sure you keep all excess mortar out of the cells where the steel vertical rods will be inserted later.

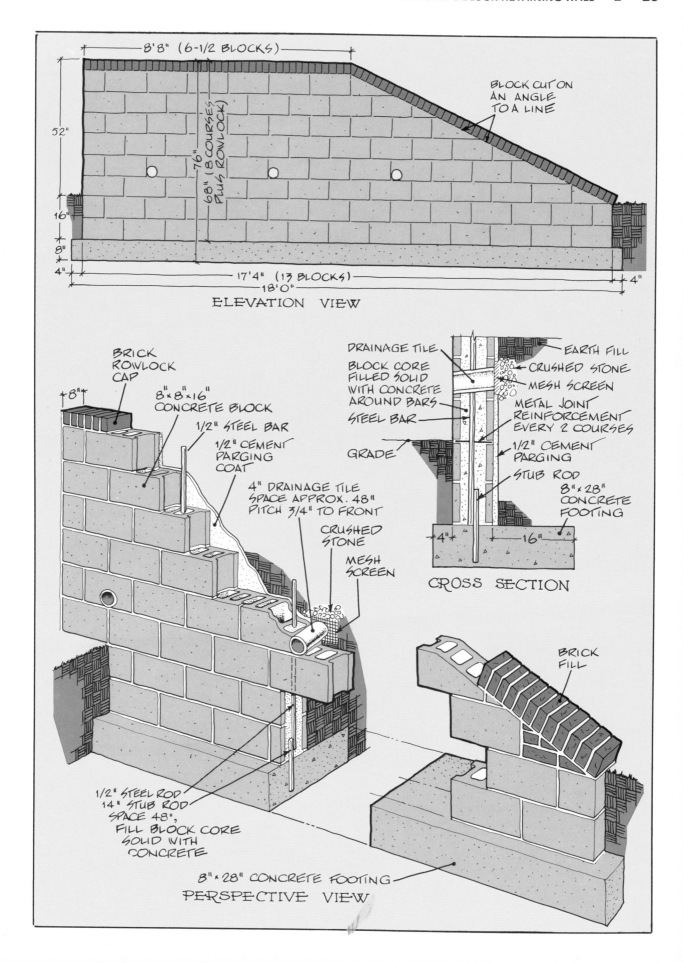

ELEVATION VIEW

8'8" (6-1/2 BLOCKS)
52"
76" (8 COURSES PLUS ROWLOCK)
68" (8 COURSES PLUS ROWLOCK)
16"
8"
4"
17'4" (13 BLOCKS)
18'0"
4"
BLOCK CUT ON AN ANGLE TO A LINE

CROSS SECTION

DRAINAGE TILE
BLOCK CORE FILLED SOLID WITH CONCRETE AROUND BARS
STEEL BAR
GRADE
EARTH FILL
CRUSHED STONE
MESH SCREEN
METAL JOINT REINFORCEMENT EVERY 2 COURSES
1/2" CEMENT PARGING
STUB ROD
8" x 28" CONCRETE FOOTING
4"
16"

PERSPECTIVE VIEW

BRICK ROWLOCK CAP
8"
8" x 8" x 16" CONCRETE BLOCK
1/2" STEEL BAR
1/2" CEMENT PARGING COAT
4" DRAINAGE TILE SPACE APPROX. 48" PITCH 3/4" TO FRONT
CRUSHED STONE
MESH SCREEN
1/2" STEEL ROD
14" STUB ROD SPACE 48", FILL BLOCK CORE SOLID WITH CONCRETE
8" x 28" CONCRETE FOOTING
BRICK FILL

Build the blockwork up three courses as shown on the plan. This will be one course above grade line. Build the drain tiles into the wall as shown. Space them about every 4' apart across the wall. Allow a slight pitch (¾") to the front of the wall to permit water to drain through. Lay a piece of aluminum screen against the drain tile on the inside of the wall and fill around the screen with gravel or crushed stone to prevent mud from clogging the wire. Cut the block or use cinder bricks and wall around the pipe securely.

As the wall progresses, every two courses of block (16") lay metal joint reinforcement in the bed joints and mortar in fully. This joint reinforcement is available at your local building supply dealer in 10' length.

Continue laying up the block wall to the height shown on the plan. At the point where the wall is cut on a rake, draw a line tightly from the low to the high point. Cut the rake (angle) on the block by measuring down from the line and marking the angle and then cutting with the brick set chisel. Lay the block on some soft earth during the cutting so that it does not crack somewhere else. Fill in with cinder bricks where it is impossible to cut the angle.

Insert the long lengths of reinforcement rods at the preestablished locations where the stub rods were placed in the footing. Fill the cells of the blocks around the rods with concrete, ramming down with a pole or board to compact. Be careful that the steel rods do not project above the top of the wall, as the brick rowlock must be laid to cap the wall. See detail on plan.

Lay the rowlock bricks on the wall to the point where it angles down. Continue laying with the aid of a line down the rake.

Tool the mortar joints as they become thumb-print-hard with either the convex (round) or V jointing tool. Brush them after they are dry.

Parge (plaster) the back of the blockwork with mortar to prevent entrance of water. Apply this in two separate coats; for best results wait overnight between coats. An additional application of a good tar waterproofing can be added to this if desired.

Let the walls cure for about a week, then fill in with earth. Tamp lightly with a shovel or tamper and let the earth settle naturally. More earth can be added as time goes by.

I would suggest waiting about 30 days to let the mortar cure before applying a waterproof paint on the front of the wall. The paint you use should contain a portland cement base, and the label should state that it does. Check with your building supplier for his recommendation on a masonry paint.

HAND TOOLS

Brick trowel	Brick set chisel
Brick hammer	Ball of line
Mason's rule	Brush
4' level	Jointing tool (convex or V)
Square	

EQUIPMENT

Mixing box	Mortarboard
Wheelbarrow	Pick and digging iron
Shovel and hoe	Rake
Bucket and hose	

MATERIALS

(*Note:* Materials given here are based on measurements of this wall. For estimating concrete or block, see the chapter on techniques of building with concrete block.)

For Footings
5 bags portland cement
1 ton sand
1 ton crushed stone

For Walls
97 blocks, 8"×8"×16"
100 standard bricks
3 drain tiles, 8" (either concrete or tile)
4 5' lengths ½" steel reinforcing rod
3 3' lengths ½" steel reinforcing rod
7 or 8 steel rods about 14" long, for stub rods in the footings
5 bags Type S masonry cement (Type N can be substituted if S is not available, but Type S is stronger.)
1½ tons building sand
3 bags portland cement, for filling rod holes in block and parging the back of the block wall
Approx. 40' metal joint reinforcement, 8" lengths, for bed joints.

9

BRICK TREE WELL

Often in the grading of home sites, it is necessary to change the existing grade in such a way that trees already on the site are above or below the new grade. Filling soil against the trunk of a tree can be disastrous to its health. Most trees are highly sensitive to changes in grade. Of course, if a tree is not needed, no special problem is created, for the tree can be removed; but on most regraded sites, there are some trees that must be preserved. If enough money is available, these trees can actually be transplanted with such large balls of earth on their roots that they will have a good chance of survival. Such an operation is costly for a large tree, and there are always several years of doubt while the braced tree is taking root and re-establishing itself. In most cases, tree wells are resorted to more often than transplanting.

A tree well is a form of a retaining wall. The purpose of any retaining wall is to hold back the earth or fill. In a case where the earth has to be filled in around a tree's roots because of a change in grade line, it becomes necessary to build a well around the roots to prevent it from dying because of a lack of water, minerals, and air. A minor fill of 8″ or less will not affect the roots, but more than this will suffocate the tree. A creative way to deal with this problem is to build a circular tree well. Because of the circular design, the earth exerts pressure more evenly and the wall acts as a key, resisting the earth's pressure much better than a straight wall would.

The smaller the circle, the more trouble you will run into keeping one brick from jutting out over the next as the circle is turned. This, of course, is if you are laying all whole bricks. For this reason, this project was built of all header-position bricks (laid crossways) with a stepped rowlock brick cap course to suit the changes in grade line. Study the plan elevation of the tree well, and you can see this.

Start this project by first establishing how big you want the circle around the tree. As for all walls that are built of masonry below grade, it will be necessary to excavate and pour a footing. Make this deeper than the frost line in your area. To deter-

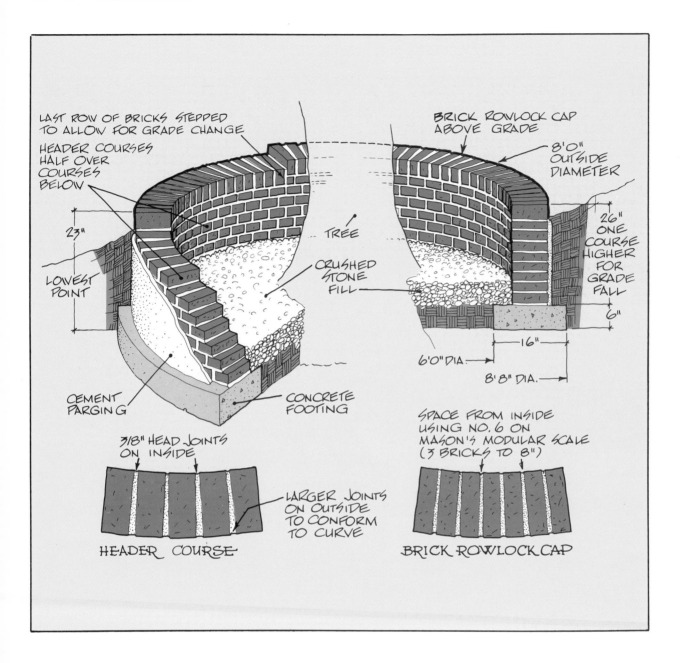

LAST ROW OF BRICKS STEPPED
TO ALLOW FOR GRADE CHANGE

HEADER COURSES
HALF OVER
COURSES
BELOW

23"

LOWEST
POINT

CEMENT
PARGING

TREE

CRUSHED
STONE
FILL

CONCRETE
FOOTING

BRICK ROWLOCK CAP
ABOVE GRADE

8'0"
OUTSIDE
DIAMETER

26"
ONE
COURSE
HIGHER
FOR
GRADE
FALL

6"

16"

6'0" DIA.

8'8" DIA.

3/8" HEAD JOINTS
ON INSIDE

LARGER JOINTS
ON OUTSIDE
TO CONFORM
TO CURVE

HEADER COURSE

SPACE FROM INSIDE
USING NO. 6 ON
MASON'S MODULAR SCALE
(3 BRICKS TO 8")

BRICK ROWLOCK CAP

mine the circle, lay a hose in a circle around the tree approximately where you want the wall to be built. Next, sprinkle a little lime around the hose to mark where excavation will take place. Excavate and pour a footing of 1 part portland cement to 2 parts sand to 4 parts crushed stone. Let the footing harden for at least one day before starting to work on it.

Lay out on the footing where the bricks will go, by again laying the hose around the circle and marking its location with a piece of chalk or lumber crayon. This wall will be 8" wide, as it is the full length of a brick.

Mix the masonry cement in a proportion of 1 part cement to 3 parts sand with enough water to suit the stiffness desired. If you want to mix a whole bag

at once, combine 16 shovels of sand with 1 bag for a good mix.

Lay in mortar a brick header course, completely around the circle, leveling and plumbing each brick with the 2' level. Fill in the mortar head joints fully to prevent any water from leaking through. Tool the inside mortar head joints with the convex or V jointer, whichever you desire, as soon as they are thumbprint-hard. See plan for layout course.

Since the wall is round, a line cannot be used to keep the wall level and plumb. I would select a number of plumb points about five bricks apart and use these as control points. However, each and every brick will have to be plumbed and leveled to assure a good job. This will entail more time than for a normal wall, but the end result will be worth the effort.

Bond each succeeding course of brick half over the one beneath as the wall is built. The inside mortar head joint will be smaller, while the joint toward the earth will be larger, since the outside of the circle must have a greater circumference than the inside.

If the ground is on a slope as shown in this project, step the last rowlock course of brick 4″ higher than the surrounding earth, to prevent earth from falling inside the well. See plan for this.

Tool or joint the top of the mortar head joints in the rowlock course flat with a slicker tool or a small pointing trowel. Brush lightly to complete the job.

Let the project set for a few days before filling in back of the wall. It is a good idea to parge (plaster) the back of the wall below grade line before filling in with earth to help make the brickwork a little more waterproof. Stop the parging at the grade line so that it does not show. To improve water drainage and keep weeds under control, fill the well with crushed stone. Level the stone and the well will be completed. Do a nice job, and this project will be the talk of the neighborhood. A few flowers planted around the outside edge of the brickwork will really set it off.

HAND TOOLS

Brick trowel
Brick hammer
Mason's rule
4′ level
2′ level

Convex or V jointer
Brush
Ball of line
Flat slicker jointer

EQUIPMENT

Mixing box
Wheelbarrow
Shovel and mortar hoe
Bucket and hose

Mortarboard
Pick and digging iron
Rake

MATERIALS
For Footing
2 bags portland cement
¼ ton sand
¼ ton crushed stone

For Wall
Approx. 375 standard bricks
3 bags masonry cement (regular)
½ ton sand

BRICK PRIVACY WALL

This project is a typical brick wall that serves as a screen affording privacy, especially to the homeowner who lives in a development or congested area. The section of wall shown on the plan can be extended or openings can be built between sections of walls as desired. The wall would be built the same no matter how long it might be.

Begin by staking off the wall and attaching a line to the stakes to mark where excavation should occur. Excavate to a depth exceeding the frost line for your area.

Mix and pour the concrete in forms or a ditch. Use a proportion of 1 part portland cement to 2 parts sand to 4 parts crushed stone or gravel. Allow at least one day for the footing to harden before attempting to build on it.

Build the concrete blockwork up to the grade line. Dry-bond the brickwork from one end of the wall to the other to determine the bond. Try to make it work in whole bricks without any cutting. This project is a double brick wall (8"), so after the outside course of brick has been laid to the line in mortar, reverse the line and lay the inside course.

Use approximately ⅜" spacing for the head and bed joints. Do not fill the middle of the wall between the bricks with mortar, as it will tend to push the wall out. Using the modular rule, lay the brick course for height to number 6 on the scale side of the rule. This will equal three courses of brick to each 8", which will work even with backing blocks when you lay them.

Study the plan for measurements and detail of construction. The best method of laying the bricks is to build a lead on each end of the wall, racking it up to a height of six brick courses. Then attach the line on each end with line blocks and fill in with brick between. Tool the mortar joints as they become thumbprint-hard.

Reverse the line on the opposite side of the wall every three courses to keep the double walls level with each other, rather than building up one side too high. It will be easier to keep the wall level and to install masonry reinforcement.

When six courses of brick have been laid, lay a length of 8" metal wall reinforcement on the wall to bond it together. Mortar this wire in place fully

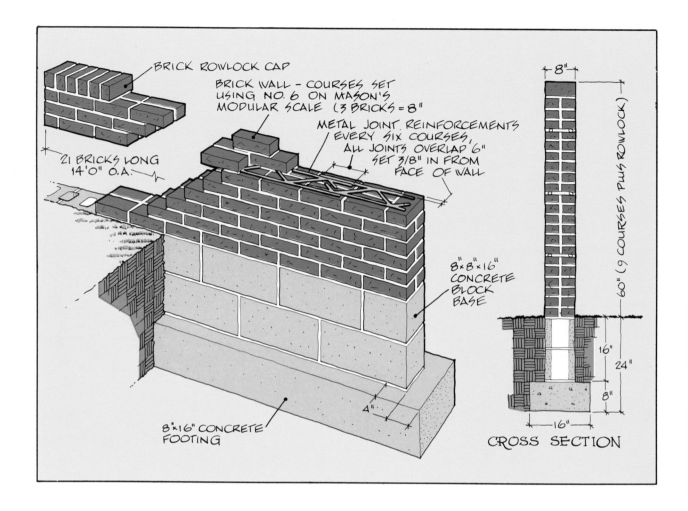

when laying the next course. Keep the wire back from the edges of all brick at least ⅜″ so the mortar joint does not pop out later.

Study the detail of how the wall reinforcement is installed in the masonry wall. Joint reinforcement is available from almost all building supply dealers in different widths and standard 10′ lengths—cut it to fit your length requirements. If laps have to be made, make sure it is lapped a minimum of 6″. The wire can be cut with regular wire or bolt cutters.

Complete the wall by laying a brick rowlock course on top and striking (jointing) the top edge with a flat joint to keep water from entering.

After the wall has set for a couple of days, fill in gently with earth around it. The brickwork can be cleaned by wetting the wall first with a hose, then scrubbing it vigorously with an acid solution of 1 part muriatic acid to 10 parts water. Use a long-handled brush, and wear eye protection and rubber gloves for safety. Muriatic acid is sold by building supply dealers and hardware stores. Avoid inhaling the fumes from the acid when mixing it in a plastic bucket. After the scrubbing is completed, rinse with plenty of clean water from a hose with a nozzle attached.

For other helpful hints on techniques of laying brick, see the chapter on techniques of laying bricks.

HAND TOOLS

Brick trowel	Ball of line
Brick hammer	Convex jointer (round)
Mason's rule	Brush
4′ level	Set of line blocks
Square	Flat slicker tool
Brick set chisel	

EQUIPMENT

Mixing box	Bucket and hose
Wheelbarrow	Mortarboard
Shovel and mortar hoe	Pick and digging iron

MATERIALS
For Footing
3 bags portland cement
½ ton sand
½ ton crushed stone

For Wall
21 concrete blocks, 8″×8″×16″
1,000 standard bricks
8 bags masonry cement (standard type)
42′ masonry joint reinforcement, 8″ lengths
1 ton sand

ASHLAR STONE PLANTER

This stone planter is perfect for that large blank space in the front or back yard. It is a combination of two different types of stone, ashlar and flagstone. Ashlar stone has a squared edge and should be laid in a consistent mortar joint thickness. It is available from many building suppliers who stock masonry materials. It comes in varying heights but usually in the same width, about 4″ as a rule. Flagstone is a flat, rather thin stone available usually in thicknesses of 1″ to 1½″ and is sold by the square foot referenced to weight. The flagstone or slate should be picked out at the building supplier for lengths of 12″. Most building suppliers will let you do this if you ask. Vermont slate is especially beautiful as it comes in red, green, and a blue shade that is natural. It is available in the eastern part of the United States at a reasonable cost. Any flagstone will work, however.

The combination of ashlar and flagstone makes for a beautiful blend in the finished planter. Planted with a variety of flowers, it will brighten the grounds of any home and requires no maintenance.

Start by staking out for the footing and excavating to a depth that will be below the frost line for your locality. Because the planter is low, a footing depth of only 6″ is needed.

After the footing has cured for a day or so, lay the block foundation up to the existing grade line, 16″ in this particular case. Use 8″×8″×16″ block, as the wall above grade line will be 4″ of stonework backed up by a 4″ block to form an 8″ wall overall. Fill the hollow core of the blocks with scrap stones and mortar to form a solid base for the stonework.

Mix the mortar a little richer for stonework. About 1 part masonry cement to 2½ parts sand will be good. Also, keep it a little stiffer than for brickwork, as stone sets more slowly because of its density and weight.

Lay a stone of fairly good size on each corner. Level and plumb it "bump to bump," as stones are not perfectly straight on the face. Attach a line from corner to corner and lay a course across. Select stones of different heights so there is a flow of mortar joints up and down. Cut the stones for

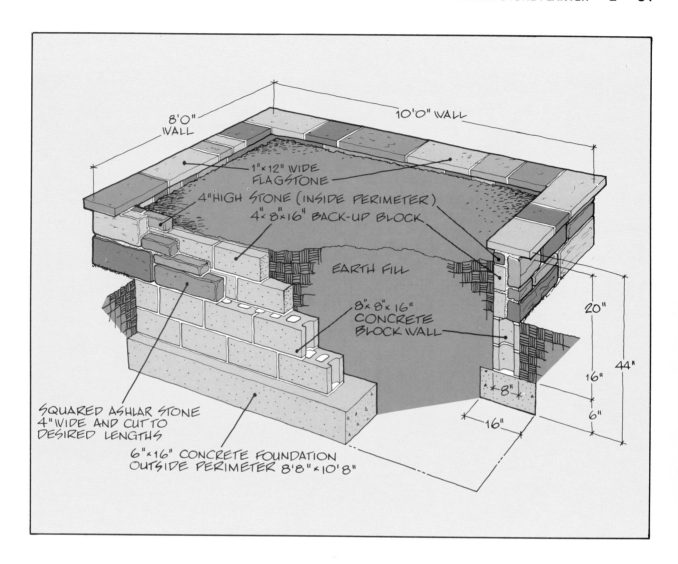

length where necessary with the wide-blade brick chisel by scoring all the way around and then returning to the front or face and striking a sharp blow. The stone should break cleanly. Occasionally lay two stones against a larger one to bond the wall together. This is known as a two-against-one and is highly desirable in stonework.

After the stonework has set a little, rake out to a depth of about ½" and smooth the mortar joints with the slicker tool. This will result in a slightly raked joint that has a smooth finish and will highlight the stones' edges. Brush the joints after they have dried enough that mortar will not smear.

Back up the inside of the planter with 4" block as shown on the cross section of wall. The remaining 4" on the inside of the wall is filled in with a course of stone, also as shown on the cross-sectional detail. Only the last stone pieces on the inside need to be tooled with the slicker, as the block will be hidden by the earthen fill required for the flowers.

Select the straightest edge of the flagstone to face the front of the wall. Try to restrict the head

and bed joint size to ½". Large mortar joints have a tendency to crack and do not look like skillful work. Divide the flagstone projection evenly over the front and back of the wall. This should be about 2" on both sides, as the wall is 8" thick and the flagstone 12" wide. If you do have to cut one for width, lay it in some sand or on soft earth and score the front and back with a chisel. Complete the cut by laying the edge to be cut off over the edge of a board and striking a sharp blow at the point where you want it to break.

Tool the mortar joints between the flagstone flat with the slicker. Brush lightly as they dry and restrike to obtain a smooth joint.

Let the stonework cure for about three days. Clean by washing down with a solution of 1 part muriatic acid to 10 parts water. Rinse off thoroughly after cleaning to remove all of the chemical solution and dirt.

I would wait for a week before filling with soil and planting any flowers to make sure that the acid solution is completely evaporated.

The possibilities for plantings to be used in the planter are very wide. If you think of the planter as a frame, the "best" plants are those most worthy of the setting. Seek out those with distinctive form, foliage, or flowers and those that have a long season of peak performance—or your personal favorites that have special meaning. Annuals, perennials, bulbs, roses, small blossoming shrubs, all-green shrubs, evergreens (which may be clipped into geometric shapes), and even trees may be placed in a planter and used with low-growing plants around them.

It is possible to keep your planter in "full" bloom all summer long by putting a succession of planted pots in them, starting with bulbs forced in pots, and then following with a succession of annuals. Every time a pot of flowers begins to fade, a replacement is brought in from the garden where a battery of pots is started and kept going for just this purpose. In the autumn potted chrysanthemums are brought in as replacements. The trick here is to use pots of the same size throughout, packing peat moss or vermiculite around them to help keep them moist and cool and watering them each day, and then putting a replacement pot in the hole left by the one removed. It can be quite a fascinating game, not to mention the interest it will add to your home. Keep in mind that the flagstone parapet is also a good place to display pots of plants. Houseplants summering outdoors can rest on the wall for a few days, and potted plants prepared especially for use around the garden can be placed along the planter wall where they will lend their beauty when it is needed and be transported to other parts of the garden when they are needed there.

HAND TOOLS

Bricklayer's trowel	Pair of line blocks
Brick hammer	Brick chisel
Heavy-duty 2-lb. striking hammer	Slicker jointer
	Pointing trowel
4' level	Brush
2' level	
Ball of line	

EQUIPMENT

Mixing box	Bucket and hose
Wheelbarrow	Mortarboard
Shovel and mortar hoe	Pick and digging iron

MATERIALS
For Footing
1 cubic yard of concrete, if you order it, *or:*
   5 bags portland cement
   1,540 lbs. sand
   1,750 lbs. crushed stone

For Wall
Approx. 45 square feet of 4"-wide ashlar stone, random heights (Estimating this at an average weight of 50 lbs. per square foot, this will be just slightly over 1 ton in weight. You will have to check this against the weight of the stone you are using in your locality.)
48 concrete blocks, 4"×8"×16", for backing up the stonework
6 bags masonry cement
¾ ton building sand
36 square feet flagstone or slate, 1"×12", for top of wall as cap

12
RUBBLESTONE PLANTER

This type of stonework is the easiest of all, because nothing has to be cut or fitted exactly. The stones are laid in the wall as they come off the rockpile. The irregularity of the bond pattern and stone shapes is what makes such stonework attractive and appealing, with its rustic natural look.

The two stone flower planters shown in the photo with the plan are simple in design—rectangular, with an average wall thickness of about 8″. The plan lists materials for only one planter, so if you wish to build a pair as shown in the photo, remember to double the amount of materials. My advice would be to build one first and then you will know pretty close how much material will be needed to build the other one.

Start by staking off where the footing line will be. Excavate to a depth safely below the frost line in your area. Pour concrete mixed to a proportion of 1 part portland cement to 2 parts sand to 4 parts crushed stone. This project takes about ¾ cubic yard of concrete for the footing. This small quantity

would be difficult to buy from a concrete company. I would suggest borrowing or renting a small utility mixer and mixing it yourself.

The plan shows two courses of concrete block for the foundation. You can use stones instead, but it would be a lot of extra work gathering them up—they're heavy. One way to acquire block inexpensively is to visit the block plant, if nearby, and buy what are called seconds. As a rule they are not weak but are sold as seconds because they have poor texture or are slightly out of square. In most cases, they can be bought for half price. You would, however, have to haul them home.

After the blocks are laid up to grade line as shown on the plan, fill the hollow cells of the block with scraps of stone or masonry fill.

Mix the mortar in the proportions of 1 part portland cement to 1 part hydrated lime to 6 parts sand. This will be a high-strength mortar, and the addition of the lime to the mix will cause it to adhere excellently to the stone. This is known in the trade

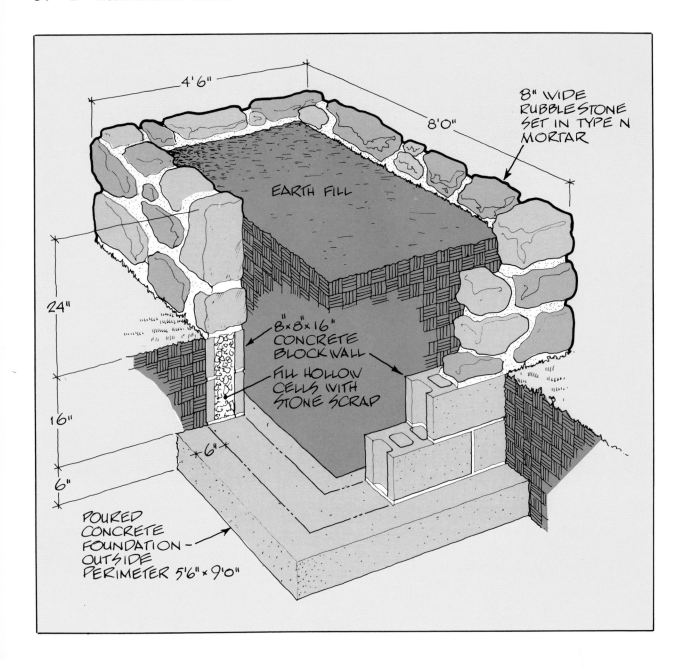

4'6"

8'0"

8" WIDE
RUBBLESTONE
SET IN TYPE N
MORTAR

EARTH FILL

24"

8"x8"x16"
CONCRETE
BLOCK WALL

FILL HOLLOW
CELLS WITH
STONE SCRAP

16"

6"

6"

POURED
CONCRETE
FOUNDATION -
OUTSIDE
PERIMETER 5'6" x 9'0"

as Type N mortar. Go easy on the water in the mix; stone mortar should be fairly stiff, because of the weight of the stone and the slow setting time.

Lay one of the larger stones on each corner in mortar as a guide. They do not have to be level and probably will be rounded on the edges anyway. Press in mortar around the edges of the stone to make sure it will be snugly in place. Attach and pull a line from corner to corner and lay the stone from one end to the other. The line will only serve as a guide for lining up the stone, not for leveling. Stones are laid from bump to bump, because of their irregularity of shape.

Fill in any big voids between the stone with smaller pieces. Fill in fully between all of the stonework with mortar, pressing in with the pointing trowel and smoothing it out. After the mortar has stiffened enough not to smear, brush it lightly to fill any pinholes and remove mortar particles.

If you wish to highlight the edges of the stone, rake out a little with a rounded piece of broom handle, resmooth with the pointing trowel, and brush. After all work has been brushed, recheck for any small hairline cracks or holes and fill them again.

Let the stonework cure for at least three days before filling the planter with earth, flowers, or shrubs. As a rule, this type of stonework is not washed down with acid solution. If you wish to remove any dry particles of mortar or dirt, just brush the surface with a wire brush. This should do the job very well.

The earth fill in the planter must provide three essentials: food, air, and water. To achieve this, an earth fill mixture should be prepared by using the following ingredients:

- *One part rich loam* (food). This is the medium for basic sustenance. It must be supplemented with fertilizer monthly during the growing season. Follow directions carefully.
- *One part vermiculite or perlite* (air). Inorganic material is needed to provide air circulation, and thus drainage, all through the root area.
- *One part peat moss* (water). Peat moss holds water, and food, in the soil. When soil shows signs of dryness, add water until it runs out bottom of container. For acid-loving plants, use straight peat moss and feed with acid fertilizer. For plants that take neutral or alkaline soil, thoroughly mix one pound of ground limestone with each bushel of peat moss.

There are many places where rubblestone planters may be effectively placed: beside the front or back doors; beneath a picture window so that plantings are lifted up making a foreground to the picture seen through the window; beside or incorporated into the terrace; flanking steps; on the upper side of a retaining wall or on the lower side with the wall as a backdrop for the plants—the possibilities seem endless.

HAND TOOLS

Brick trowel	Brush
Brick hammer	Short length (about 6") of
Heavy-duty mason's	old broom handle for
hammer	raking out mortar
4' level	joints

Ball of line	Mason's rule
Several masonry chisels	Pointing trowel
2' square	

EQUIPMENT

Mixing box	Bucket and hose
Wheelbarrow	Mortarboard
Shovel and hoe	Pick and digging iron

MATERIALS
For Footing

3 bags portland cement
1,100 lbs. sand
1,300 lbs. crushed stone

For Wall

Approx. 2 pickup loads (½-ton truck rubblestone) or about 2 tons by weight (*Note:* This may seem like a lot, but rubblestone weighs about 125 lbs. per cubic foot. You will have to check the weight of the stone you are going to buy and work with that figure. If the stones are free, there is no need to worry about the weight, only your time to haul it.)

4 bags portland cement
4 bags mason's hydrated lime
1 ton sand

(*Note:* The amount of mortar you will need for the wall depends on the percentage of voids or filling in around the stone. It is impossible to estimate accurately how much mortar will be used. The best practice is to buy about half of the mortar materials listed here, use them, and estimate the balance needed based on how far that went. Professional estimators who figure stonework always allow about 15% more than really is needed. The handyman should not, as there will be too much waste and you cannot return it.)

13
RAISED BRICK AND
STONE TREE WELL

ometimes when the grade has to be cut away for a patio area or a similar change is being made, the roots of a tree will become exposed, and this in time will kill the tree. This is the reverse of the problem that occurs when it is necessary to fill in around a tree, but the consequences are the same.

The answer is to build a raised retaining wall around the tree to protect the roots. It will, however, in all probability be necessary to water the tree, especially during a long dry spell, because the tree roots are above the existing grade and will not receive enough moisture just from rainfall. Filling with peat moss around the top of the well will help to keep the soil moist.

Since the brick wall will be laid in mortar, you will have to excavate below the existing frost depth and install a footing. Mix the concrete in a ratio of 1 part portland cement to 2 parts sand to 4 parts crushed stone and water. A footing 6" in depth will be more than enough for this project.

After the footing has cured for at least one day, lay a hose around the footing at the approximate position that you want the wall built. Mark this

circle with a piece of chalk or crayon. As mentioned in the materials list, you can use brick from the footing up, which will establish the correct curvature and bond, or you can lay block or old chipped bricks under the grade. The choice is up to you. Remember, however, to lay out the brickwork at least one course below the grade line so the blocks do not show.

Lay the bricks in mortar bed joints, but *do not* put any mortar head joints (mortar on the ends of the bricks) on bricks. This is done so moisture will weep out through the open head joints, thereby helping to prevent cracking of the wall. Plumb and level each brick as you build your way around the circular wall. If you wish, a curved template of plywood can be made for a portion of the circle and be held from one plumb point to another. This will make the job go a little faster. After each course of brick is laid, take a nail or piece of wire and clean out any of the head joints that might have become filled with mortar.

Tool the mortar bed joints as they become thumbprint-hard and brush when dry. Lay metal

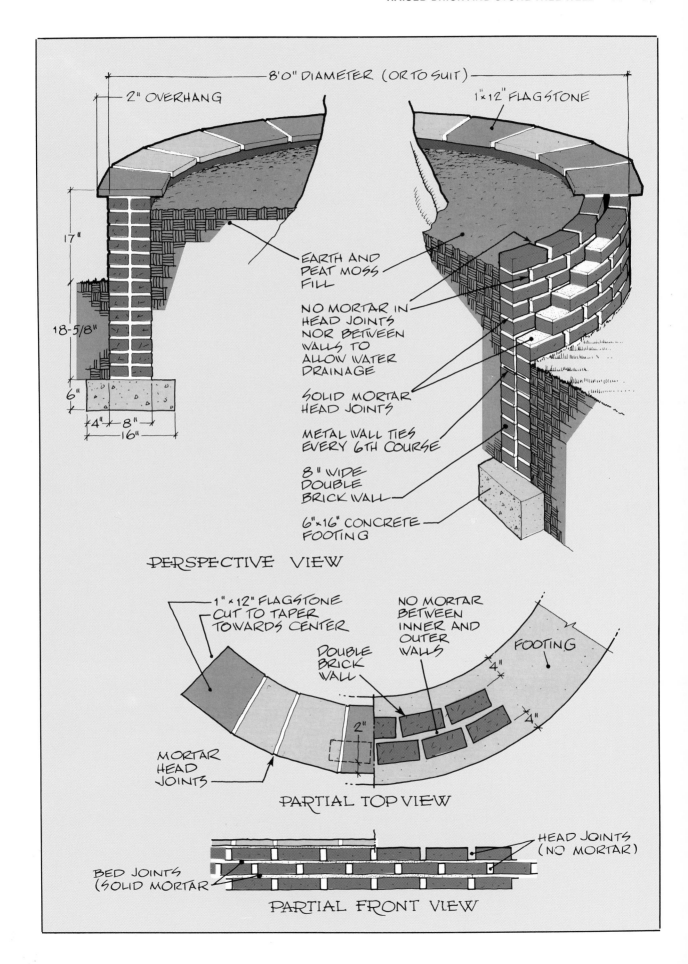

8'0" DIAMETER (OR TO SUIT)

2" OVERHANG

1"×12" FLAGSTONE

17"

18-5/8"

6"

4" 8"
16"

EARTH AND PEAT MOSS FILL

NO MORTAR IN HEAD JOINTS NOR BETWEEN WALLS TO ALLOW WATER DRAINAGE

SOLID MORTAR HEAD JOINTS

METAL WALL TIES EVERY 6TH COURSE

8" WIDE DOUBLE BRICK WALL

6"×16" CONCRETE FOOTING

PERSPECTIVE VIEW

1"×12" FLAGSTONE CUT TO TAPER TOWARDS CENTER

NO MORTAR BETWEEN INNER AND OUTER WALLS

FOOTING

4"

4"

DOUBLE BRICK WALL

2"

MORTAR HEAD JOINTS

PARTIAL TOP VIEW

HEAD JOINTS (NO MORTAR)

BED JOINTS (SOLID MORTAR

PARTIAL FRONT VIEW

wall ties across the double wall every sixth course as shown on the plan to bond the walls together. These ties are available from building supply dealers and are known as "brick veneer corrugated wall ties." Mortar them in well so that they are effective.

You will have to plumb the outside of the wall and the inside, as every course should be approximately 8" wide at the top (total width of both walls). Do not fill in the center of the walls with mortar.

Lay the flagstone cap, setting it out 2" on the front and back of the wall. This should work out perfectly if the brick double wall is kept to 8" in width and you purchase 12"-long flagstone. These figures can, of course, vary some, but to keep the wall in balance should be adhered to as closely as possible. The flagstone used in the photo is Vermont slate and comes in several natural colors. It is especially attractive.

Since the wall is built on a circle and the flagstones are rectangular, you will have to cut a little off the edges on an angle to avoid large mortar joints at the outer edge of the wall. This can be accomplished by laying the flagstone flat on the ground or sand, marking the part to be cut off, and chipping along the edge with the brick hammer. If a flagstone has to be cut completely in two, score a line on one side with the hammer and chisel, then turn it over and score the other side. Then lay the stone over a piece of pipe or 2×4 in line with the scored edge and strike a sharp blow with the hammer and chisel. The stone should break cleanly at the desired place. Point up the top of the mortar joint between each stone by using a flat slicker tool or pointing trowel.

After the project has been completed, run a piece of wire or nail between each of the head joints in the brickwork to remove any particles left and to ensure good drainage. Wait at least a week, then clean the brickwork and stonework with a solution of 1 part muriatic acid to 10 parts water.

Fill with earth and peat moss inside the tree well to finish the project.

HAND TOOLS

Brick trowel	Brick chisel
Brick hammer	Convex jointer
Mason's rule	Brush
4' level	Slicker jointing tool

EQUIPMENT

Mixing box	Bucket and hose
Wheelbarrow	Mortarboard
Shovel and hoe	Pick and shovel

MATERIALS

For Footing

2 bags portland cement
¼ ton sand
¼ ton crushed stone

For Wall

Approx. 1,000 bricks if bricks are used from the footing up to the finished height. (*Note:* Block could be substituted or old chipped bricks used below the grade line.)

Approx. 28 square feet of flagstone, for top. (*Note:* Flagstone should be about 12" long, but width could vary, as shown in project. Also, I would try to pick stones that are about the same thickness, which is 1".)

7 bags masonry cement
1 ton building sand
24 wall ties

14

RADIAL BRICK FLOWER BED

Any home can be enhanced by building this beautiful brick flower bed. The combination of the two right-angle walls along with the radial quarter-circle front wall presents a striking appearance when blooming with lovely summer flowers. A rakeout joint in the brickwork and a clever header and rowlock cap really finish off this project.

Start by marking off where the footing will be excavated. Do this by staking two right-angle level lines in each direction. Dry-bond 16½ bricks on top of the ground in each direction to determine length of footings.

Next, to lay out the front radius wall, attach another line to the corner point where the two walls meet, and swing an arc from one end of the wall to the other. Mark the outline of the walls with powdered lime to excavate by.

Excavate for the footing, making sure it is below the freeze line in your area. (Footing should be excavated at least twice the width of the finished wall.)

Mix the concrete for the footing to a ratio of 1 part portland cement to 2 parts sand to 4 parts crushed stone, with enough water to blend the mix together. Since the project is rather large, it will take just about 1 cubic yard of concrete. This should not take too long to mix if you can get a small utility mixer and one of the neighbors to stop by for a while. You could also order 1 yard from a concrete company, but it would be expensive.

Allow the footing to cure a day or so before starting to do any further work.

You will notice that in the plan I have shown all brick from the footing to the top of the project. This was done because of the curvature of the front wall. If you do not wish to lay brick below the grade, concrete block can be substituted instead.

To lay out the brickwork, draw a circular line on the footing with a lumber crayon by attaching a line to a nail driven at the point where the two straight walls form a corner and then swinging in an arc from the end of one straight wall to the end of the

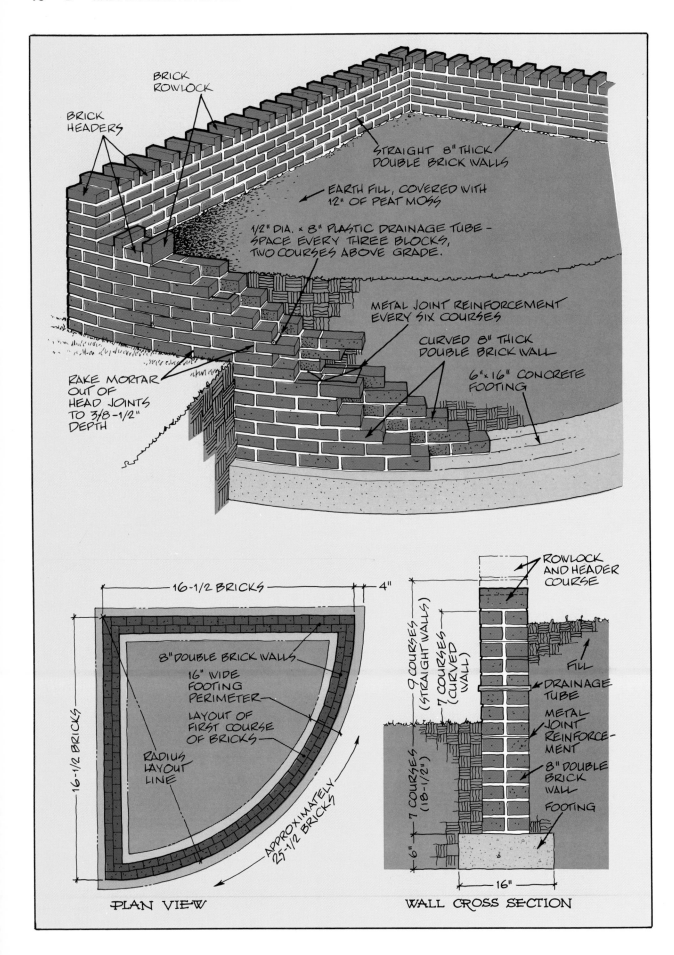

BRICK
HEADERS

BRICK
ROWLOCK

STRAIGHT 8" THICK
DOUBLE BRICK WALLS

EARTH FILL, COVERED WITH
12" OF PEAT MOSS

1/2" DIA. × 8" PLASTIC DRAINAGE TUBE—
SPACE EVERY THREE BLOCKS,
TWO COURSES ABOVE GRADE.

METAL JOINT REINFORCEMENT
EVERY SIX COURSES

CURVED 8" THICK
DOUBLE BRICK WALL

6"×16" CONCRETE
FOOTING

RAKE MORTAR
OUT OF
HEAD JOINTS
TO 3/8-1/2"
DEPTH

16-1/2 BRICKS

4"

8" DOUBLE BRICK WALLS

16" WIDE
FOOTING
PERIMETER

LAYOUT OF
FIRST COURSE
OF BRICKS

16-1/2 BRICKS

RADIUS
LAYOUT
LINE

APPROXIMATELY
25-1/2 BRICKS

PLAN VIEW

ROWLOCK
AND HEADER
COURSE

9 COURSES
(STRAIGHT WALLS)

7 COURSES
(CURVED WALL)

FILL

DRAINAGE
TUBE

METAL
JOINT
REINFORCE-
MENT

8" DOUBLE
BRICK
WALL

7 COURSES
(18-1/2")

FOOTING

6"

16"

WALL CROSS SECTION

other straight wall. You can determine where the ends of the straight walls are by striking a chalk line for the two straight walls on the footing.

Then dry-bond the 16½ bricks in each direction. Lay a brick in mortar at each corner point, attach a line, and lay the first course. In laying out the radius front wall, a line cannot be used, because of the curvature. Instead, slide the 4′ level along as you lay out the radius to keep the wall level. Plumb every brick on the face.

Build the masonry work up to the grade line, plumbing each brick on the radius wall. If you're using block for the foundation, start laying out the bricks two courses below the finished grade to work out the bond correctly. Lay metal joint reinforcement in the wall every six courses of brick for strength and to bond the wall together.

Build the two straight walls ahead of the curved front, each course interlocking the brick at the ends into the curved wall. Insert in the front wall on the second course, every third brick, a short length of ½″ plastic tubing in the head joint. This will help to drain excessive water that may build up in the flower bed and prevent cracking. See detail on plan.

Rake out the mortar joints on the front of the wall to a depth of approximately ⅜″ with a rakeout jointer with a nail in it. Brush when dry enough. This will create texture and appearance of depth in the brickwork. Rakeout jointers are available at your local building supply dealer and come with skate wheels attached to assist movement. The nail is adjustable to suit different depths of joints. Also, a nail driven into a straight length of wood could be used to run along the face of the wall. However, I much prefer using the factory made rakeout jointer described, which is not expensive.

Lay the last course (cap) on the wall by alternating brick headers and rowlock-position brick as shown in the photograph and on the plan. Level each one front to back and attach a line on the wall to keep them straight.

Point the mortar joint on all of the top flat work with a slicker tool and pointing trowel. It is very important that there be no holes for water to enter, as it will cause the joints to crack after a period of time.

Let the masonry cure for about a week and then wash down with a solution of 1 part muriatic acid to 10 parts water, making sure that the wall is pre-wetted first and rinsed down thoroughly after the washing to remove all acid solution. This cleaning will bring out the color of the bricks and accent your work. Wait another couple of days before filling with earth and doing any planting. When the planter is filled with colorful flowers and pleasing shrubs, it will be a wonderful accent piece for your home.

HAND TOOLS

Brick trowel	Rakeout jointer
Brick hammer	Slicker jointer
Mason's rule	Ball of line
4′ level	Pointing trowel
2′ level	Line blocks
Brick chisel	Brush

EQUIPMENT

Mixing box	Mortarboard
Wheelbarrow	Pick and shovel
Shovel and hoe	Wire cutters
Bucket and hose	

MATERIALS

For Footing

5 bags portland cement
1,500 lbs. sand
1,750 lbs. crushed stone

For Wall

1,700 standard bricks, if bricks are used from footing to top of wall (*Note:* Blocks could be substituted for masonry under grade.)
13 bags masonry cement
1½ ton building sand
78 lineal feet masonry steel joint reinforcement, 8″ lengths (available in 10′ lengths)
70″ plastic or vinyl tube, ½″ diameter (available from most hardware stores)

15
ONE-FLUE BLOCK CHIMNEY

This project is a real necessity and money saver in these times of high energy costs. Many persons are adding a one-flue chimney for wood stoves. Some very important things to remember when building a one-flue chimney are:

- Form well-filled mortar joints.
- Keep the blockwork plumb and level, for appearance and to keep the weight of the chimney balanced.
- Insert and fasten steel anchors to the chimney and into the studding of the house next to chimney, every 24" in height, to prevent any movement.
- Build the top of the chimney at least 2' above the peak of the roof for a good draft.
- If using chimney block, as this project does, make sure that a space of 1" is left between the block and the exterior framed wall of the house for safety.

- You will have to get a permit and stand inspection to keep your home insurance in force.

There are other things to consider, but the ones listed above are especially important to remember.

Start out by excavating for the footing. Make sure that it is below the frost line for your area. Also, make sure that the chimney is not being built on filled earth. If there are any questions at all, excavate down to the original footing line of the house and install your footing there. The building inspector will check on this.

Since a chimney is a concentrated weight in a small area, the footing has to be deeper than for a normal wall. I recommend pouring concrete at least 12" in depth and making the footing at least 12" wider than the chimney at all points around it. This will spread the load more evenly over an area. Reinforcement rods of steel or some type of metal should be put in the concrete when it is poured. In

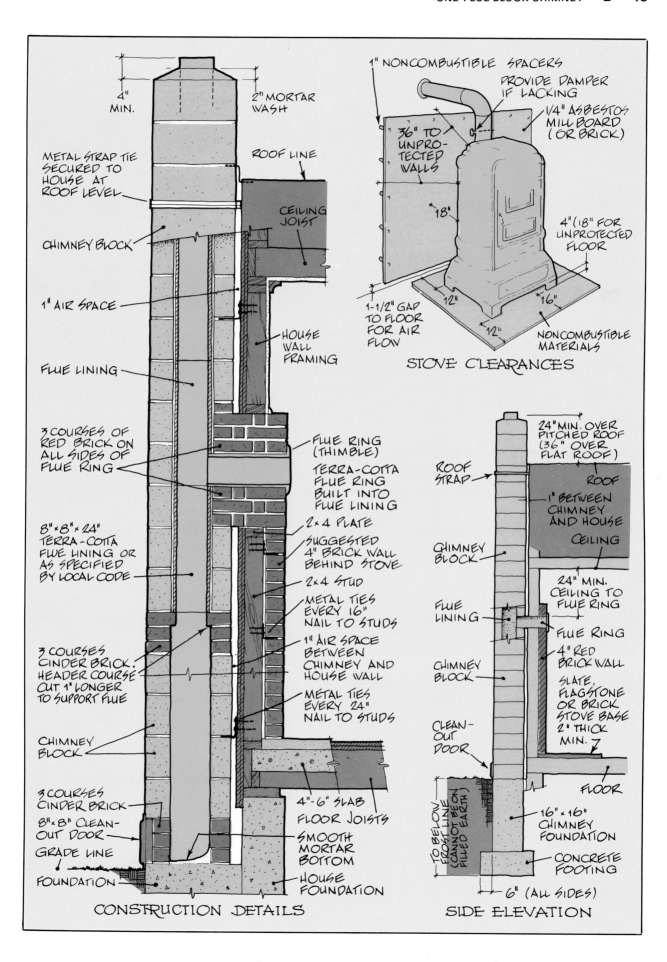

4" MIN.

2" MORTAR WASH

METAL STRAP TIE SECURED TO HOUSE AT ROOF LEVEL

ROOF LINE

CEILING JOIST

CHIMNEY BLOCK

1" AIR SPACE

HOUSE WALL FRAMING

FLUE LINING

3 COURSES OF RED BRICK ON ALL SIDES OF FLUE RING

FLUE RING (THIMBLE)

TERRA-COTTA FLUE RING BUILT INTO FLUE LINING

2 x 4 PLATE

SUGGESTED 4" BRICK WALL BEHIND STOVE

2 x 4 STUD

8"x8"x24" TERRA-COTTA FLUE LINING OR AS SPECIFIED BY LOCAL CODE

METAL TIES EVERY 16" NAIL TO STUDS

1" AIR SPACE BETWEEN CHIMNEY AND HOUSE WALL

3 COURSES CINDER BRICK: HEADER COURSE CUT 1" LONGER TO SUPPORT FLUE

METAL TIES EVERY 24" NAIL TO STUDS

CHIMNEY BLOCK

3 COURSES CINDER BRICK

4"-6" SLAB

FLOOR JOISTS

8"x8" CLEAN-OUT DOOR

GRADE LINE

SMOOTH MORTAR BOTTOM

FOUNDATION

HOUSE FOUNDATION

CONSTRUCTION DETAILS

1" NONCOMBUSTIBLE SPACERS

PROVIDE DAMPER IF LACKING

1/4" ASBESTOS MILL BOARD (OR BRICK)

36" TO UNPROTECTED WALLS

18"

4" (18" FOR UNPROTECTED FLOOR

1-1/2" GAP TO FLOOR FOR AIR FLOW

12"

12"

16"

NONCOMBUSTIBLE MATERIALS

STOVE CLEARANCES

24" MIN. OVER PITCHED ROOF (36" OVER FLAT ROOF)

ROOF STRAP

ROOF

1" BETWEEN CHIMNEY AND HOUSE

CEILING

CHIMNEY BLOCK

24" MIN. CEILING TO FLUE RING

FLUE LINING

FLUE RING

4" RED BRICK WALL

CHIMNEY BLOCK

SLATE, FLAGSTONE OR BRICK STOVE BASE 2" THICK MIN.

CLEAN-OUT DOOR

FLOOR

TO BELOW FROST LINE (CANNOT BE ON FILLED EARTH)

16"x16" CHIMNEY FOUNDATION

CONCRETE FOOTING

6" (ALL SIDES)

SIDE ELEVATION

many localities, the inspector will want to look at the excavation before you pour the footing to make sure that it is the correct size and there is reinforcement in it. Be sure to check this out before pouring the concrete.

Mix the concrete in a ratio of 1 part portland cement to 2 parts sand to 4 parts stone, with water. Let the footing cure at least two days before starting to build on it.

Chimneys can be built of brick or chimney blocks. This project is built of chimney block. Chimney blocks are available from building supply dealers in different sizes. For the standard 8″×8″ flue that is used for a wood stove chimney, a 16″×16″ block is used. See illustration of chimney block with plan. The flue lining is set inside the chimney block to make it fireproof. Flue linings are made of a burned clay called terra-cotta and will withstand very high temperatures. They are sold in either round or rectangular shapes. The standard flue liner used is rectangular and comes in 24″ lengths or sections.

Mix masonry cement mortar in a proportion of 1 part masonry cement to 3 parts sand. A half batch can be made by mixing ½ bag masonry cement to 8 shovelfuls of sand or a full batch by using 1 bag masonry cement to 16 shovelfuls of sand.

Build the cleanout door into the outside face of the chimney, using solid cinder bricks in place of the chimney block, because they can be fitted to the door better. I recommend using a cast-iron cleanout door, as it will resist rust. The cleanout door and solid cinder bricks are available from your building supply dealer. Smooth a sloping coat of mortar inside the bottom of the chimney at the point where the cleanout will be installed to make it easier to remove soot.

Continue laying the chimney blocks, keeping a space of 1″ from the house wall for fire protection to a height of 24″ below where the flue ring (thimble) will be installed.

Using cinder bricks again, rack out on the inside of the chimney approximately 1½″ to serve as a shelf for the first flue lining to sit on. See details on plan.

Lay red bricks on the inside of the house, starting at a point 8″ under where the flue ring is to be inserted, up to where it will be set in place. Most building codes call for at least 8″ of red brick to surround all flue rings for fire protection. The red bricks will rest on a 2×4 plate built into the studded house wall for strength. If this chimney is being added to an existing house, a hole is cut through the house wall at this point to hook up the chimney to the stove. Special flue linings are available that have a round hole made in them for the purpose of inserting a flue ring for hooking up a stovepipe. If these linings are not available in your locality, the hole can be cut by packing the flue tightly with sand, then marking a circle on the outside of the flue where it should be cut. Cut by punching a hole in the center of the circle with a point chisel, working outward from the center by chipping with the head of the hammer. Continue chipping until you can fit the flue ring snugly into it. See illustration.

Build the flue ring into the flue liner, keeping it flush with the inside edge and mortaring neatly to obtain a tight, smooth fit. Build three courses of red brick under, over, and around the flue ring to complete the installation. See detail of flue ring installation on plan.

Continue building the chimney up to the roof, laying metal wall ties in the chimney and nailing into the studding every 24″. Set and mortar in succeeding flue linings as the chimney progresses.

Metal flashing will have to be cut and inserted in the mortar joint slightly above the roofline. I would advise getting a carpenter or a roofer to do this for you to ensure a leakproof job. Contact him beforehand so he is on the job with the materials when needed.

Build the chimney at least 2′ above the ridge of the roof for a good draft and apply a mortar cove on top. It is a good idea to protect the roof and prevent staining and damaging shingles by laying a tarpaulin or sheet of heavy plastic on it.

Make sure that a heavy-duty metal strap anchor is securely installed around the chimney near the edge of the roof and lagged into the framework of the roof. The chimney should cure for a period of 30 days before subjecting it to any heat.

Last, do not attach a TV antenna to the chimney, as it is not designed for this purpose. In high winds an antenna can whip about enough to shake a chimney loose.

A draft check can be made by holding a lighted piece of paper at the end of the flue ring. The smoke should draw freely into the ring and up the chimney.

Some Important Rules One Should Follow When Installing a Wood Stove

1. The stove should be positioned on a noncombustible material such as brick or an approved floor protection material.
2. The stove should be inspected to make sure it has no cracks or broken places that would make it unsafe.
3. A layer of sand or firebrick should be in the bottom of the stove to keep in from burning out.

4. Floor protection should extend out 6″ to 12″ from the sides and back of the stove and 18″ from the front where the wood is loaded.

5. The stove should be spaced a minimum of 36″ away from combustible materials. If not, fire-resistant materials must be used to protect all woodwork or materials that could burn.

6. Stovepipe should be of 22-gauge or 24-gauge metal.

7. Stovepipe should not be reduced between the stove and chimney flue.

8. Your insurance company should be notified that a stove has been installed or your policy could be canceled or insurance invalid.

9. A building or fire inspector should approve the final installation. (This is required in many localities by law.)

10. Hot ashes should not be removed and carried out in an unprotected container over a carpeted floor.

11. An old chimney should be cleaned before a stove is hooked up to it. Many local fire departments will clean and inspect an old chimney for a donation as a public service.

12. Stovepipe should not pass through a floor or closet on the way to the flue.

13. The flue ring (thimble) should be of a fireproof material such as terra-cotta tile and should be the same size as the stovepipe for a neat fit.

14. Stovepipe should never pass directly into the flue lining, as it may burn out.

15. Stovepipe should enter the chimney through a flue ring that is higher than the outlet of the stove.

16. The stovepipe should have a little slope (minimum 1″) from where it enters the flue ring to the stove.

17. There should always be at least 18″ between the top of the stovepipe and the ceiling of the room; 24″ would be better.

18. Total length of a stovepipe should be less than 10′.

19. A damper should be installed in stovepipe near the stove even if the stove is an airtight design, to control excessive draft on a windy day.

20. Clean your stove and chimney at least once a year. Professional chimney cleaners should be listed in your local phone directory.

HAND TOOLS

Brick trowel	Brick chisel
Brick hammer	Point chisel
Mason's rule	Convex jointing tool
4′ level	Brush
2′ square	

EQUIPMENT

Mixing box	Bucket and hose
Wheelbarrow	Scaffolding
Shovel and hoe	Rope and pulley

MATERIALS

(*Note:* The amount of mortar, chimney block, etc., will depend on how high your individual chimney is, so no amounts are given here.)

1 cleanout door (of cast iron)

1 flue ring (thimble) either 6″ or 8″ diameter to fit stove outlet (*Note:* Length of flue ring depends on distance from inside of house to flue.)

8″×8″ flue linings, 2′ high each

Standard red bricks, approx. 150 if you build brick in back of stove; approx. 30 if you only build them around flue ring)

40 cinder bricks

Masonry cement

Sand

Crushed stone for footing mix

1 bag portland cement, for footing mix

Chimney blocks (*Note:* Each block is 8″ high; number needed will depend on height of your chimney.)

Metal strap ties to be nailed into the studding of house

16

RUBBLESTONE FIREPLACE

Concerning materials for this project, the ones listed here are for the building of the fireplace from the first-floor line to the ceiling line and therefore do not include all of the materials needed to build the entire chimney. Since the chimney is of typical construction, I felt it was not necessary to repeat the information given in the previous chimney project.

When building a rubblestone fireplace such as this—in fact, any kind of fireplace—always consult your local building code. Some codes do not permit certain types of fireplace construction. Should local codes be ignored and fire results, you may be in trouble with local authorities, as well as your insurance company. Also, pay strict attention to basic construction rules given here. If a fireplace is not carefully built, it may develop smoke leaks and serious structural faults. Poor operation and home damage are the possible consequences.

This quaint stone fireplace utilizes the beauty and charm of natural stonework. There is no specif-

ic pattern or bond to follow; the stones should merely be fitted and laid on their best edge for stability and be pointed up neatly. You can vary the length of the fireplace to suit your needs for room size. Since the average wall width is 6″, it would be a good idea to dry-fit all stones in position before actually bedding them in mortar. It saves taking them back down and recutting if there is a projecting bump or knob on the back edge.

After the foundation has been built to the floor level, lay out the fireplace as shown on layout plan, making sure that the firebox is of the proper size.

Mix your mortar for stonework a little stiffer than for brickwork, to a proportion of 1 part portland cement to 1 part hydrated lime to 6 parts sand, with water.

Build the stonework in the front and sides up to about 16″, leaving the opening for the firebox. Then mark the center of the opening with a square and lay the firebrick hearth as shown in the photograph. Complete the firebox by laying out and lay-

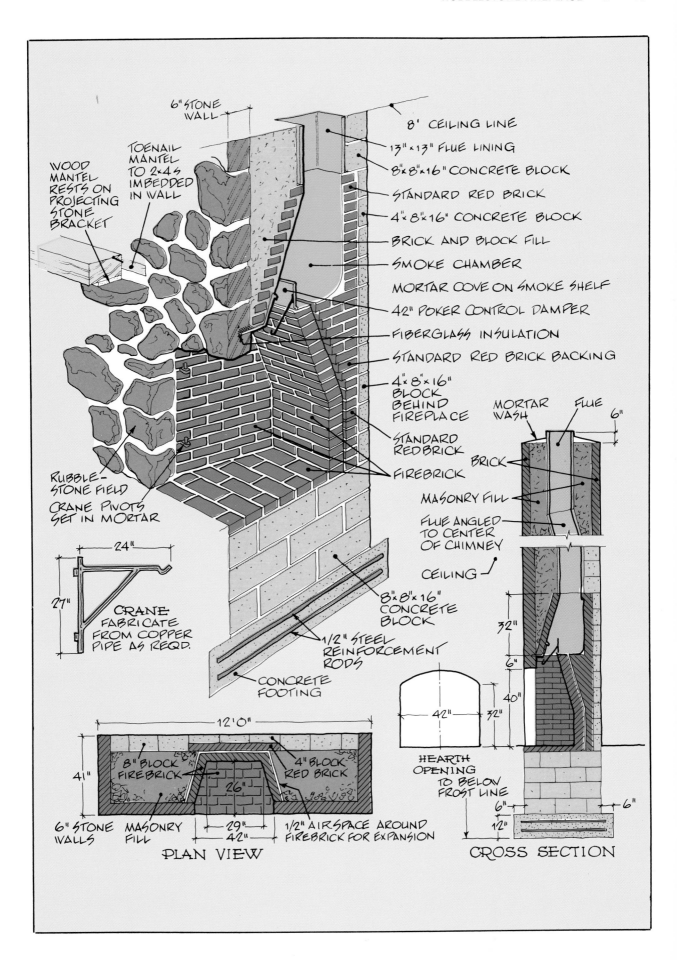

6" STONE WALL

TOENAIL MANTEL TO 2×4s IMBEDDED IN WALL

WOOD MANTEL RESTS ON PROJECTING STONE BRACKET

8' CEILING LINE

13"×13" FLUE LINING

8"×8"×16" CONCRETE BLOCK

STANDARD RED BRICK

4"×8"×16" CONCRETE BLOCK

BRICK AND BLOCK FILL

SMOKE CHAMBER

MORTAR COVE ON SMOKE SHELF

42" POKER CONTROL DAMPER

FIBERGLASS INSULATION

STANDARD RED BRICK BACKING

4"×8"×16" BLOCK BEHIND FIREPLACE

STANDARD RED BRICK

FIREBRICK

RUBBLE-STONE FIELD

CRANE PIVOTS SET IN MORTAR

MORTAR WASH

FLUE

6"

BRICK

MASONRY FILL

FLUE ANGLED TO CENTER OF CHIMNEY

CEILING

24"

27"

CRANE FABRICATE FROM COPPER PIPE AS REQD.

8"×8"×16" CONCRETE BLOCK

1/2" STEEL REINFORCEMENT RODS

CONCRETE FOOTING

32"

6"

40"

12'0"

41"

8" BLOCK FIREBRICK

4" BLOCK RED BRICK

26"

29"
42"

1/2" AIRSPACE AROUND FIREBRICK FOR EXPANSION

6" STONE WALLS

MASONRY FILL

PLAN VIEW

HEARTH OPENING TO BELOW FROST LINE

42"

32"

6"

6"

12"

CROSS SECTION

ing the sides and back up as shown on the plan.

Continue building the stonework up to the height where the arch starts. Set the damper in position on a bed of mortar on top of the firebox. This is shown on the plan.

Fill in and around the firebox with bricks, blocks, and mortar as the work progresses, being careful to leave approximately ½" expansion space around the firebox.

Build and set in place a wood arch form in the front of the firebox to support the stonework over the opening. See plan for measurements.

Place some fiberglass insulation around the damper on all edges, as mortar should not touch the metal. This is done to allow room for expansion of the metal damper when it gets hot.

Tool the mortar joints inside the firebox area with a convex (round) jointer.

Tool the finished mortar joints of the stonework by rubbing out slightly with a length of old broom handle and then brushing the joints lightly when they are dry enough not to smear.

Continue building the fireplace up to the height where the stone brackets are projected out for the mantel. The finished mantel height is about 54", as shown on the plan. Inset a few 2×4 wood blocks behind where the mantel will rest so that you can screw or toenail on the back edge of the mantel to the fireplace later.

Corbel (rack) the brickwork in the smoke chamber area to the point where the flue lining will rest. This is shown on the plan and is about 32". These proportions are very important to maintain if the fireplace is going to operate correctly. Don't vary them much!

Set the flue lining in place as shown on the plan. The stonework is then built up to the 8' ceiling height, filling in the inside of the fireplace as you work with bricks, blocks, and mortar.

After the stonework cures for several days, it can be cleaned by washing with a solution of 1 part muriatic acid to 10 parts water. Make sure that you wet the walls first and flush them down well after with clean water to remove any chemicals. It is also a good idea to wear safety eye protection when cleaning.

If the room is occupied as a living space and this fireplace is an improvement, then use a stiff bristle brush and vacuum off excess dust.

The crane and copper kettle, if you use them, should not be attached to the firebrick until after a week has passed to allow the mortar a chance to harden. In addition, a fire should not be built in the fireplace for 30 days.

The crane and copper kettle, if you use them, should not be attached to the firebrick until after a week has passed to allow the mortar a chance to harden. In addition, do not build a roaring fire in your rubblestone masonry fireplace right after it is finished. Allow 30 days to elapse for seasoning of the mortar; otherwise serious cracks may develop.

When selecting fuel for your fireplace, real wood logs are generally most acceptable, though some home owners may prefer, on occasion, to use coal or charcoal. At any rate, a wood fire is usually required to start other fuels.

Logs may be split or whole and should be 16" to 22" in length. Dry, seasoned hardwoods (such as hickory, oak, walnut, or most available fruitwood) make the best fuel. Soft woods burn away too quickly; wet, green woods should be avoided because they smolder instead of burning clean. Also avoid the use of scrap lumber or refuse. This material, especially when excessively dry, produces a great many sparks that escape up the flue and become a fire hazard.

Keep your fireplace screen in place or closed, especially when leaving the room, as sparks might pop out unobserved to smolder on the floor or carpet. Always remember to OPEN the damper when starting a fire and CLOSE the damper when the fire is completely out.

(*Note:* This rubblestone fireplace was built on the first floor of the house in a family recreation room. The handyman who built it did not build any outside hearth in front of the firebox but laid carpet right up to the fireplace. I recommend strongly that an outer hearth of stone be built for extra fire protection in place of the carpet shown in the photograph.)

HAND TOOLS

Bricklayer's trowel	Chalk box
Bricklayer's hammer	Square
Mason's folding rule	Flat slicker tool
4' level	Brush
Pointing trowel	Convex jointer
Broad-edge chisel	Heavy hammer (2-lb. size
Ball of line	a good choice)
Line pin and nail	

EQUIPMENT

Wheelbarrow	Bucket and hose
Shovel and hoe	Mortarboard
Mixing box	

MATERIALS
For Footing and Foundation
2½ cubic yards concrete

Approx. 8 pieces of steel reinforcement rod, ½" diameter and 50" long, placed approximately 2' on center to reinforce the footings

Concrete blocks (*Note:* The number needed will vary depending on whether you are building over a full basement or a crawl-space foundation or on a ground-level slab. Estimate the number of blocks by multiplying the total lineal feet around the foundation by .75. This will give you the number of blocks in one course. To find the total number of courses high, divide 8″ into the total height expressed in inches.)

Mortar (*Note:* Estimate quantity by allowing 1 bag masonry cement to every 28 blocks.)

Sand (*Note:* Each 8 bags masonry cement require 1 ton sand.)

For Fireplace and Chimney

(*Note:* Quantities are for fireplace and chimney in the photograph, from the first floor to the ceiling only.)

Approx. 150 square feet rubblestone (*Note:* This is for walls 6″ wide and will be about 4 tons of stone. This figure includes the front and sides of the fireplace up to 8′ height as shown on the plan.)

12 bags portland cement
12 bags hydrated lime
2 tons sand
90 concrete blocks, 4″×8″×16″, for back of chimney
20 concrete blocks, 8″×8″×16″, for last two courses on back wall

Standard firebrick (*Note:* If the firebricks are laid on their widest edge as shown in the photograph, it will take about 167. Since firebricks are about 55¢ each, this is costly. You can lay them on their narrowest edge and it will take about 80. Either way is acceptable if cost is a factor.)

500 standard red bricks, for building in and around the firebox and interior of the fireplace

1 poker control damper, 42″ long

Flue linings, 13″×13″ (*Note:* The number needed will depend on how high you build your chimney. Figure one flue section for every 2′ in height.)

1 wood beam or timber to serve as a mantel (*Note:* If you so desire, the beam can be run the entire length of the fireplace front. The handyman who built this fireplace selected a mantel about 10′ long. Old barn timbers, if not decayed or full of bugs, make good mantels.)

Accessories as desired—(The copper kettle was obtained at an auction and the crane made by the handyman.)

Masonry fill (*Note:* The bulk of this fireplace and chimney will require a lot of masonry fill, which could be expensive if all new materials are used. I would suggest using old used bricks, or second-run blocks that may have irregularities but are sound to fill in with. These are available from block plants in your area at about half of the new cost of blocks.)

17

BRICK WALL WITH ARCHED OPENING IN KITCHEN

A firm but stable foundation must be built under the brick walls. If you are building on a concrete slab on grade, it will safely support the weight with no problems. However, if you are remodeling your kitchen, either a steel beam must be inserted under where the walls are to be built to carry the load or an 8" concrete block wall must be built up from the foundation floor. This project description assumes that you will make provisions for a firm, stable base under the walls and deals only with the masonry work from the first floor to the ceiling line.

Start by laying the brick walls out as shown on the layout plan and dry-bonding brick around the wall areas to make it work in whole bricks if possible. The foundation or base should be built beforehand as mentioned earlier.

Since the bricks are used and will vary in size, it can be expected that you will have to cut one here and there to make it work out.

Before laying out the openings for the stoves or ranges, measure the units you are going to install to make sure they are going to fit. The measurements shown on the layout plan are only an average size.

Lay the brick courses to number 6 on the course-counter spacing rule. Unless the bricks run extra-thick, this should work for joint spacing. This will be approximately 8¼" for every three courses of bricks laid.

Every six courses in height, build metal wall ties into the mortar joints to bond the two thicknesses of wall together. See plan illustration of wall ties in wall.

Tool the mortar joints with a rakeout jointer approximately ½" deep and brush out clean after they have dried enough not to smear.

If you are building an oven and microwave into the brickwork as the photograph shows, lay out the opening and when the height for the top of the opening is reached cover it with two 3½"×3½" angle

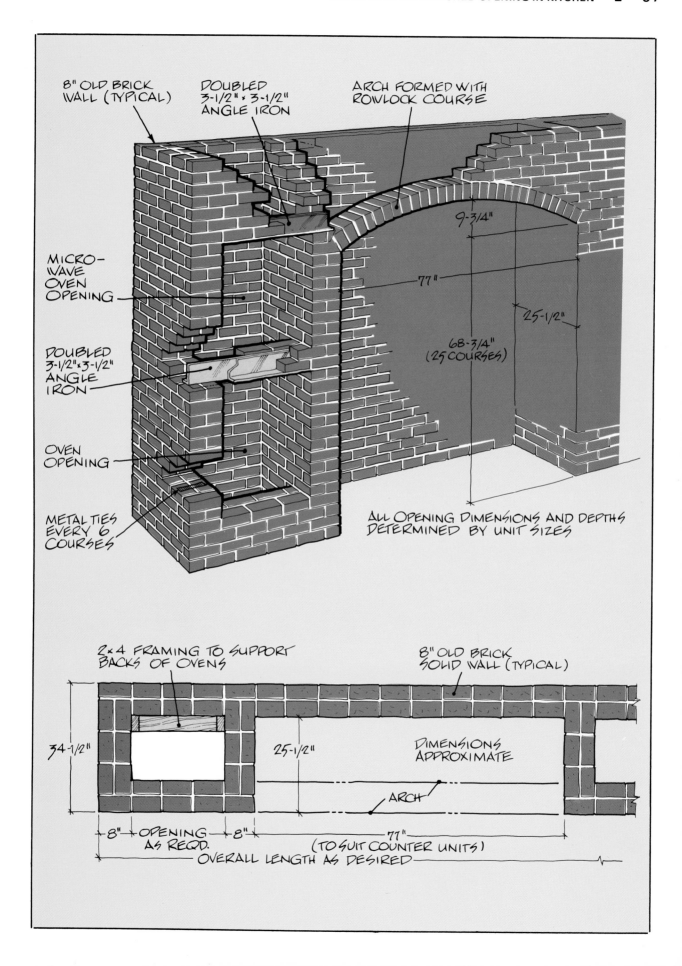

8" OLD BRICK WALL (TYPICAL)

DOUBLED 3-1/2" × 3-1/2" ANGLE IRON

ARCH FORMED WITH ROWLOCK COURSE

MICRO-WAVE OVEN OPENING

DOUBLED 3-1/2" × 3-1/2" ANGLE IRON

OVEN OPENING

METAL TIES EVERY 6 COURSES

9-3/4"

77"

25-1/2"

68-3/4" (25 COURSES)

ALL OPENING DIMENSIONS AND DEPTHS DETERMINED BY UNIT SIZES

2×4 FRAMING TO SUPPORT BACKS OF OVENS

8" OLD BRICK SOLID WALL (TYPICAL)

34-1/2"

25-1/2"

DIMENSIONS APPROXIMATE

ARCH

8" OPENING AS REQD. 8"

77"

(TO SUIT COUNTER UNITS)

OVERALL LENGTH AS DESIRED

irons laid back to back. The angle irons should be long enough so that at least 6" rest on both sides of the opening for good support.

Later when the oven is installed in the brick wall, it may be necessary to build a 2×4 studded frame to support the rear of the ovens. The electrical wiring should also be built in behind the brickwork to conceal it. You may want to contact a registered electrician about this before starting the brickwork, as he will probably want to install a receptacle of some type.

Continue laying the brickwork up to the point where the arch starts. Build a wood form using 2×4s for legs and brace it into position securely. The dimensions for this are on the plan. Mark off the individual brick spacing around the edge of the wood form with a pencil and the spacing rule. The arch shown in the photograph was a rowlock and a segmental arch. Lay the brick arch on the form, making sure that the mortar head joints are well filled. Do not lay any mortar bed joint on the wood form, as it would have to be chiseled out later.

The brickwork above the arch on the inside of the wall should also be tooled by raking out the mortar joints and brushing. If you wish, a drop ceiling could be installed later to seal the inside area off. The choice is up to you!

Continue laying the brickwork up to the 8' ceiling line. After the arch has cured for at least two days, remove the form gently and repoint up on the bottom side with fresh mortar. You may have to take a small chisel and cut out any mortar joints that have flushed out too far.

The brickwork should be cleaned if possible by washing down with a solution of 1 part muriatic acid to 10 parts water. Be sure to wet the wall first so the solution does not soak in too far and burn the brickwork. Rinse after cleaning with water through a hose nozzle to remove all of the chemicals. If you don't want to use muriatic acid, most building suppliers stock other chemical cleaners you can substitute. In any event, it is a good idea to wear safety eye protection during this process.

In the project shown, the cabinets and appliances were built into the opening and appropriate trim fitted around the edges to make a neat fit. The electric service wiring was passed through holes drilled in the floor in back of the units.

A length of 1" copper pipe was installed slightly below the crown of the arch on the inside of the wall to hang pots and pans on. If you want to install an exhaust fan over the appliances, this could also be built into the ceiling on the inside of the arched opening. This is optional, but should be considered before building any type of ceiling inside the recessed opening.

HAND TOOLS

Bricklayer's trowel	2' square
Bricklayer's hammer	Chalk line
Mason's folding course-counter rule	Line pin and nail
	Rakeout jointer
4' level	(skatewheel)
Brick chisel (broad-edge)	Pointing trowel
Ball of line	Brush

EQUIPMENT

Mixing box	Hose
Shovel and hoe	Mortarboard
5-gal. bucket	Wheelbarrow

(*Note:* In addition, you will need a few basic carpentry tools to construct the arch form and some 2×4 braces and lumber for the arch form.)

MATERIALS

(*Note:* The bricks utilized in this project were old, used ones. If you use the same type, make sure that they are sound and not soft. You can detect this by tapping them lightly with a hammer. If they have a solid ring to them, then they are okay to use. Old bricks are also slightly larger than modern standard bricks. The number of bricks needed will depend on how long you build your kitchen wall. For this reason, no specific amounts of materials are stated; methods of estimating the bricks are given instead.)

Bricks (*Note:* Used bricks are a little larger than modern standard; it will take approximately six bricks per square foot of wall area. Since the walls are 8" wide (two bricks), be sure to remember to double your square footage amounts.)

Mortar (*Note:* Use regular masonry cement mortar mixed in the proportions of 1 part masonry cement to 3 parts sand, with water. For larger batches, mix either 8 shovels sand to ½ bag masonry cement or 16 shovels sand to 1 full 70-lb. bag masonry cement. Each bag of mixed masonry cement mortar will lay 125 bricks.)

Sand (*Note:* Allow 1 ton sand to every 8 bags mortar.)

4 angle irons, laid back to back to cover the openings over the oven and microwave where they are built into the brickwork, two for each opening

Wood form, to support the arched opening (*Note:* Rise and length of arch is indicated on front elevation view. Size of your bricks could vary this measurement slightly.)

Metal wall ties, to be built into the mortar joints every six courses of bricks to bond the walls together as shown on the plan.

A few lengths of 2×4, to support the back of the oven and microwave (*Note:* Frame these in to suit your individual needs.)

18
BRICK SCREEN WALL

Photograph of project courtesy of Brick Institute of America.

This attractive brick screen wall is the ideal solution for a carport where you want to hide trash cans, lawn tables, etc. It also will provide somewhat of a windbreak for your car. The building of the holes in the panel will take more time than laying up a regular brick wall, and close attention must be paid that the halves line up true vertically for the best appearance. This type of brickwork is also known as latticework. Check your local building codes to make sure that they will allow this type of wall to be built. It saves a lot of trouble later.

Start your construction by excavating for the wall below the existing frost line for your area. Mix the concrete in a proportion of 1 part portland cement to 2 parts sand to 4 parts crushed stone, with water.

Pour the concrete level in a trench, leaving the top untroweled so that there will be a good bond between the mortar and concrete when the first course is laid out.

After the footing has set for at least one day, dry-bond, then lay out the first course of bricks level and plumb, in mortar as shown on the plan.

Form about ⅜″ head joints. You will notice that the wall is brick from the footing up. If you want to substitute 4″ concrete block, they will work equally well up to the grade.

Space all brick coursing height to number 6 on the modular rule scale.

Lay the brickwork up to the finished grade line, using line blocks on each corner and filling in to the line between. Build the return ends as you lay up the corner, making sure that the return is square with the front wall. Use the 2′ framing square periodically to keep the wall square.

At grade line, dry-bond from one end of the wall to the other the screen or lattice pattern, using a 6″ cut on the corner and spacing the halves of bricks as shown on the plan. These are identified by letters on repeating screen bond layout view on plan. It is a very simple pattern but has to be laid plumb with the brick below as the wall is built to achieve a neat repeating pattern.

Lay in mortar the 6″ cut brick on each end of the wall, attach a line, and lay out the first course as shown. This is done by laying a cut half in position

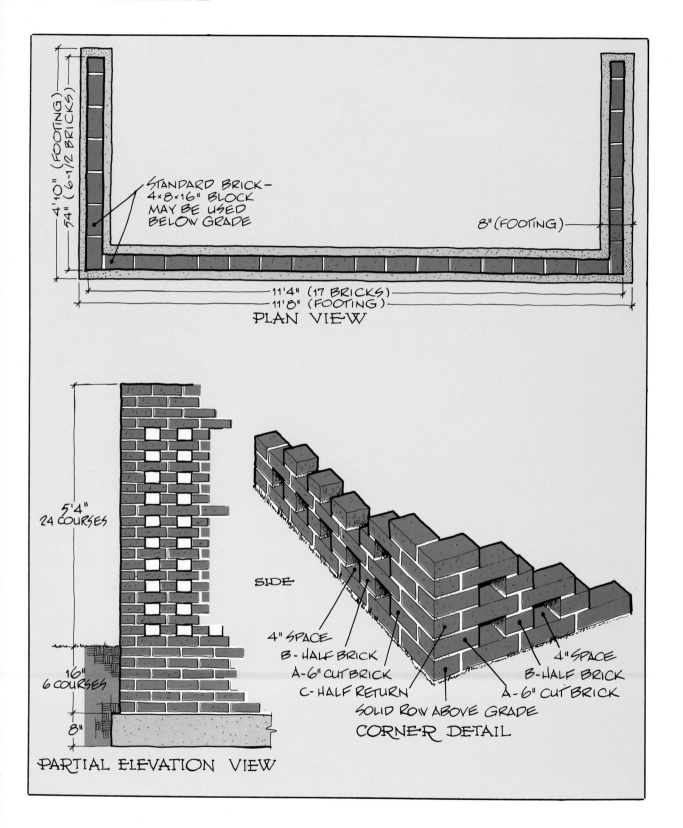

STANDARD BRICK—
4×8×16" BLOCK
MAY BE USED
BELOW GRADE

4'0" (FOOTING)
54" (6-1/2 BRICKS)

8" (FOOTING)

11'4" (17 BRICKS)
11'8" (FOOTING)

PLAN VIEW

5'4"
24 COURSES

16"
6 COURSES

8"

PARTIAL ELEVATION VIEW

SIDE

4" SPACE
B- HALF BRICK
A- 6" CUT BRICK
C- HALF RETURN
SOLID ROW ABOVE GRADE

4" SPACE
B- HALF BRICK
A- 6" CUT BRICK

CORNER DETAIL

centered on the joint where the two bricks come together below. See B on the layout detail of repeating screen bond. The cut part of the half brick will not show, as it is laid with the cut end to the back of the wall. However, try to make the cut as neat as possible. Start the second course with a half return brick on the corner (shown as C on drawing) followed by a whole brick spanning over to the middle of the cut half brick on the course below. Continue laying the full bricks for the rest of the course as shown on the plan and in the photograph. Repeat the first pattern described after the full brick course is laid in place.

Using a slicker (flat-bladed tool), compress the mortar bed joints in fully around all the pigeonholes, especially on the jamb or sides to stabilize the bricks better. The head and bed joints facing the exterior should be tooled to match the mortar joint finish on your house or wall adjoining the project.

Check the vertical alignment of the halves at least every three courses to make sure they do not shift one way or the other. Do this with the 4' level.

Continue building the screen pattern wall up to the height where the full brick course covers the panel, ending the screen design. A great deal of care should be exercised during the bricklaying to avoid putting too much pressure against the wall, as it could be easily pushed over or the set of the mortar joints broken. Remember that this is only a 4" wide wall (single-brick width), and until it sets will be relatively weak in a vertical position. After the mortar has cured, it will be more than strong enough for what it is designed to do.

After the wall is completed, wait about a week and wash down with a solution of 1 part muriatic acid to 10 parts water. Be sure to wet down the wall first before scrubbing and rinse well afterward with a hose with a nozzle attached. All protruding mortar joints on the inside of the wall should have been cut off flush with the trowel as the wall was built and any holes filled at that time.

This project will require more patience and skill than a regular bricklaying job, because of the intricate design. Study the plans well before starting and during the construction, and follow directions stated above. The end result will be very pleasing and a testimony to your bricklaying skills.

HAND TOOLS

Bricklayer's trowel	Slicker tool
Bricklayer's hammer	Ball of line
Mason's modular rule	2 line blocks
4' level	Brush
2' level	2' square
Brick chisel (wide-bladed)	Small square
Pointing trowel	Pencil
Convex jointer	

EQUIPMENT

Mixing box	Bucket and hose
Wheelbarrow	Mortarboard
Shovel and hoe	Pick and digging iron

MATERIALS
For Footing
3 bags portland cement
700 lbs. sand
Approx. 900 lbs. crushed stone

For Wall
Approx. 900 standard bricks, including the halves you will have to cut for the panel screen. (*Note:* This is allowing about 40 bricks for breakage. You may want to just buy 1,000 bricks as this would allow you plenty for breakage and they are usually cheaper by the thousand.)
6 bags masonry cement (regular)
1 ton building sand

19

BRICK PAVING LAID IN MORTAR

When selecting brick for paving work, always pick a solid, hard brick, as it will hold up longer with less chance of cracking. Such brick is classified as SW (severe weathering) by the supplier. Regular paving brick is available in three different sizes. The most popular and considered the standard size is shown as A in the illustration. B is a little smaller in width and length. C is wider but only 1⅝″ thick and is useful where depth is a problem.

There are several different bond patterns one can select for paving work. The simplest is the "running bond," which consists of all whole brick, laid one half lap against the one next to it. A half brick is required every other course at the starting point to form the half-lap bond.

A popular pattern arrangement is the "basketweave" bond. This pattern is a little more difficult because of differences in brick sizes. It is accomplished by laying two bricks in the same position next to each other and then reversing the next two to form a weave pattern. There are several variations of the basketweave possible for different appearances.

Last, the "herringbone" bond is the most intricate but is very beautiful when laid correctly. It is simply a running bond, but with the bricks laid on a 45° angle to the border. It does require more cutting than the others, as the starting 45° angle point pieces have to be laid against the border to form the bond pattern. This bond is used more for mortarless paving, as maintaining of the joints is not as critical as when laying in mortar.

Fewer problems will be encountered if you stick to one of the more simple patterns such as the running or basketweave bond for your first effort. The skills of buttering mortar on and laying the brick in mortar can be complicated if the bond is intricate. The illustration shows some of the most often used paving bonds for bricks.

Laying the Brick Paving

Laying brick paving in a mortar bed requires a concrete base. Start by excavating the soil to the required depth. Allow approximately 4″ for crushed stone or gravel, 4″ for concrete slab, and 3″ for the brick and mortar bed joint. This is a total of 11″ in

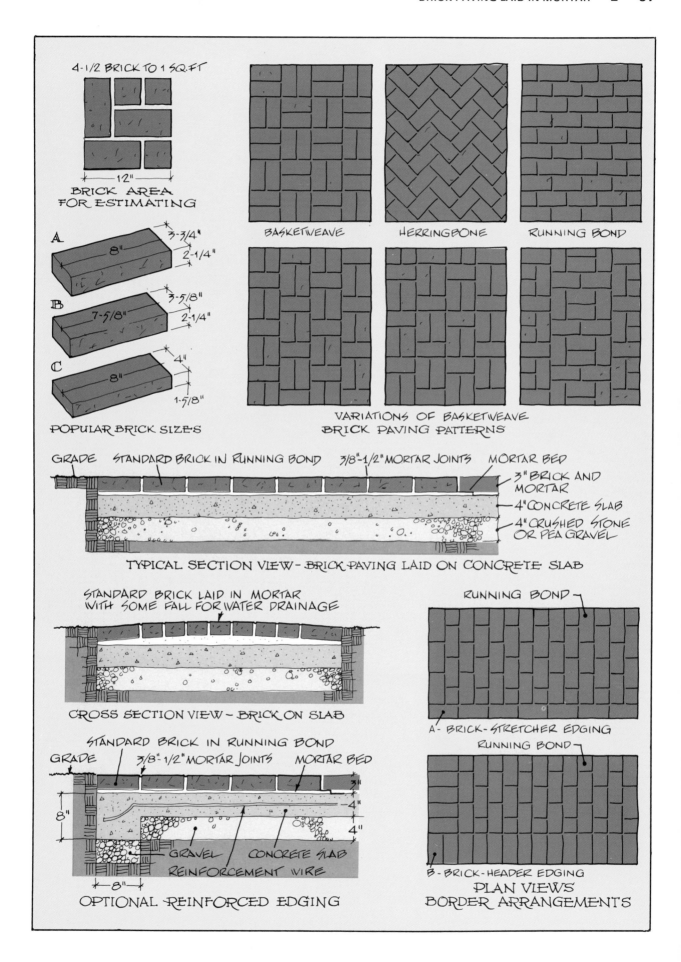

4-1/2 BRICK TO 1 SQ.FT

12"

BRICK AREA FOR ESTIMATING

BASKETWEAVE · HERRINGBONE · RUNNING BOND

A — 8" — 3-3/4" — 2-1/4"

B — 7-5/8" — 3-5/8" — 2-1/4"

C — 8" — 4" — 1-5/8"

POPULAR BRICK SIZES

VARIATIONS OF BASKETWEAVE BRICK PAVING PATTERNS

GRADE — STANDARD BRICK IN RUNNING BOND — 3/8"-1/2" MORTAR JOINTS — MORTAR BED

3" BRICK AND MORTAR
4" CONCRETE SLAB
4" CRUSHED STONE OR PEA GRAVEL

TYPICAL SECTION VIEW – BRICK PAVING LAID ON CONCRETE SLAB

STANDARD BRICK LAID IN MORTAR WITH SOME FALL FOR WATER DRAINAGE

CROSS SECTION VIEW – BRICK ON SLAB

RUNNING BOND

A - BRICK-STRETCHER EDGING

STANDARD BRICK IN RUNNING BOND
GRADE — 3/8"-1/2" MORTAR JOINTS — MORTAR BED
8"
3"
4"
4"
GRAVEL — CONCRETE SLAB
REINFORCEMENT WIRE
8"

OPTIONAL REINFORCED EDGING

RUNNING BOND

B - BRICK-HEADER EDGING

PLAN VIEWS BORDER ARRANGEMENTS

depth. See illustration of standard brick laid in running bond.

When digging out the soil, try not to take out any more than needed, as it will have to be filled back in with stone. A simple method of establishing this depth is to drive a wood stake in the ground along the edge of the area to be excavated and attach a line. If you want the line to be level, hang a line level on it. If there is a fall or pitch, as in most walks, just attach the line up from the existing grade the same height at both ends. The finished brick surface when completed should be slightly above the existing grade line. If a brick walk is to be laid that has no slope lengthwise, then the middle should be slightly crowned to drain the walk. See illustration.

Tamp the earth solidly after excavating to form a good base. Fill in with about 4″ of crushed stone or gravel and pour 4″ of concrete, screeding off even with a wood straightedge. Some old fence or concrete reinforcement wire will help if this is imbedded in the concrete for extra strength. Reinforcement wire is available from your local building supplier.

If you live in an area where severe cold weather occurs, it is a good idea to pour the edge of the concrete base deeper for more protection and strength. This can be accomplished by pouring it 8″ by 8″ along the edges. This is shown in the illustration.

The biggest problem area in a walk or patio is along the edges, where the bricks tend to work loose. A practical method of preventing this is to lay the border in a different position than the rest of the paving. This also creates a pleasing appearance, as the border would be a different pattern than the rest of the work. The illustration shows two possibilities, a brick stretcher along the edge and a brick header along the edge.

To lay the brick paving in mortar, I suggest first laying the border course to the line to the desired height, making sure that you fill all of the mortar joints fully. Then stretch a line across the walk or patio and lay the brick to the line in the opposite direction. Butter each brick with mortar, using the trowel, and settle it in place by tapping with the trowel or hammer handle until it is firmly set. As the

Brick paving laid in mortar on a concrete slab surround this beautiful pool.

mortar becomes thumbprint-hard (so that the indentation of your thumb will remain in the mortar when pressed in), tool the surface of the joints. A concave joint (half-round) can be used, but I prefer to tool them flat with a slicker or a small pointing trowel. After the mortar has dried so it won't smear, brush lightly and restrike if necessary.

Two good tips to remember when laying brick paving are not to rake out the mortar joints and repoint later but tool them as you go (this will achieve a much better bond, and there will be less trouble with mortar discoloration), and, if the work is being done in hot dry weather and the bricks are of the sand type, which are apt to soak up moisture quickly, dampen the bricks with a fine spray of water from a hose so they do not set too rapidly in the mortar joints. If the bricks are very hard and smooth they will not have to be wetted.

Mix your mortar the same stiffness as when laying brick in a regular wall. You should be able to settle the brick into the mortar bed without excessive pounding with the trowel or hammer handle. A rubber crutch tip slipped on the end of your trowel handle will help to prevent cracking the brick. Never lay cracked bricks or bricks that have a hollow sound, as they will in all probability soon break.

After the work has been completed and has set for at least four or five days, it can be cleaned of smears or mortar splatters by scrubbing with a solution of 1 part muriatic acid to 10 parts water, making sure that you flush off all acid solution and dirt after the scrubbing process. The result will be a beautiful, long-lasting, maintenance-free brick pavement you can be proud of for many years to come.

HAND TOOLS

Bricklayer's trowel	Pair of line blocks
Bricklayer's hammer	Slicker tool
Mason's modular rule	Small square
4' level	Pencil
Line level	Brush
Brick chisel (wide-bladed)	Knee pads or a piece of
Pointing trowel	foam rubber
Ball of line	

EQUIPMENT

Mixing box	Mortarboard
Wheelbarrow	Pick and digging iron
Shovel and hoe	Straightedge of 2×6 wood
Bucket and hose	about 10' long

MATERIALS

(*Note:* No attempt is made here to estimate the amount of materials for a specific brick paving job, because no two walks or patios are the same size. The information below will adapt to fit your particular situation. Always allow a little more than you really need for waste.)

● If the bricks are laid with the widest side facing up, which is the usual method for paving, there are 4½ standard bricks for every square foot of area. See illustration.

● One bag of masonry cement mixed with 16 shovels of sand will lay approximately 115 bricks. This includes the bed and head joint allowances. (Brick paving requires a little more mortar than when laying up a wall, because of the differences in the concrete base.) As a rule, the average mortar head joint between brick laid in paving work is about ½".

● Brick paving is laid on a concrete base or slab usually 4" in thickness and reinforced with concrete reinforcement wire for extra strength. If you have some fence wire lying around the place, this will work in place of the reinforcement (wire should be in good condition).

● Concrete is estimated by calculating the number of square feet and converting it into cubic feet, then to cubic yards, which can be ordered from your local concrete company, or you can mix it if the amount is not too great. If all of this is a bit too complicated for you, there are handy slide rules available from most companies that supply concrete. Also, the concrete supplier will figure the amount you need, if you specify the size of the slab.

● Crushed stone, cinders, gravel, or stone screenings (stone dust) all will work fine as a base under the slab. Generally, about 4" is enough to level the base area for concrete. Crushed stone and the other materials mentioned for fill are sold by weight, usually by the ½ ton or ton, and are inexpensive. Again, if you tell the building material supplier the size of the slab area, he can tell you how much stone will be needed.

20

MORTARLESS BRICK PAVING

The primary advantage of laying paving bricks without mortar is that there are no mortar joints to crack from expansion and contraction of the earth from changing weather conditions and freezing and thawing. In addition, it is less expensive, as there are fewer materials used, and the time involved to do the work is shorter.

Start by excavating to provide a base. The normal depth allowance is about 8", which consists of 4" of crushed stone, 1" to 2" of stone dust or sand, and the thickness of the bricks, which is about 2¼". The thickness of the tar paper is minor and need not be considered. The finished brick paving should be slightly (about ½") above the finished grade and have some slope to run water off.

The ground should be firm. Brick laid on recently filled dirt will naturally have a tendency to sink in spots. One nice thing, however, is that the bricks can easily be taken up and relaid, adding a little more sand or stone dust. If you had laid bricks in a mortar bed, this would be a major chore.

When the excavation is completed, spread approximately 4" of stone or gravel over the area to serve as a base.

There are two frequent methods used for laying a border for brick paving. First is to pour a concrete base along the border area measuring 8" by 8". The first-course or border bricks are laid in mortar in this base. The theory is that this will help to prevent the border from shifting and provide a firm edge to work against. This method is shown in the illustration.

The second and most often used method is to lay the border course in a different position than the rest of the walk or patio and place it deeper in the earth for greater resistance to shifting or movement. This method is also shown in the illustration.

Borders can be laid in regular running bond as long as the earth is filled in tightly against the edges to prevent it from moving. Once the border is established, a line can be set up across the project to lay bricks to, making the work a lot easier and more accurate.

After the border is laid in place, spread a 1" to 2" layer of sand or stone dust over the area to be paved. This can be simplified by cutting a screed board from a piece of 1×6 wood and notching the ends out to allow for brick paving. Drive a stake and

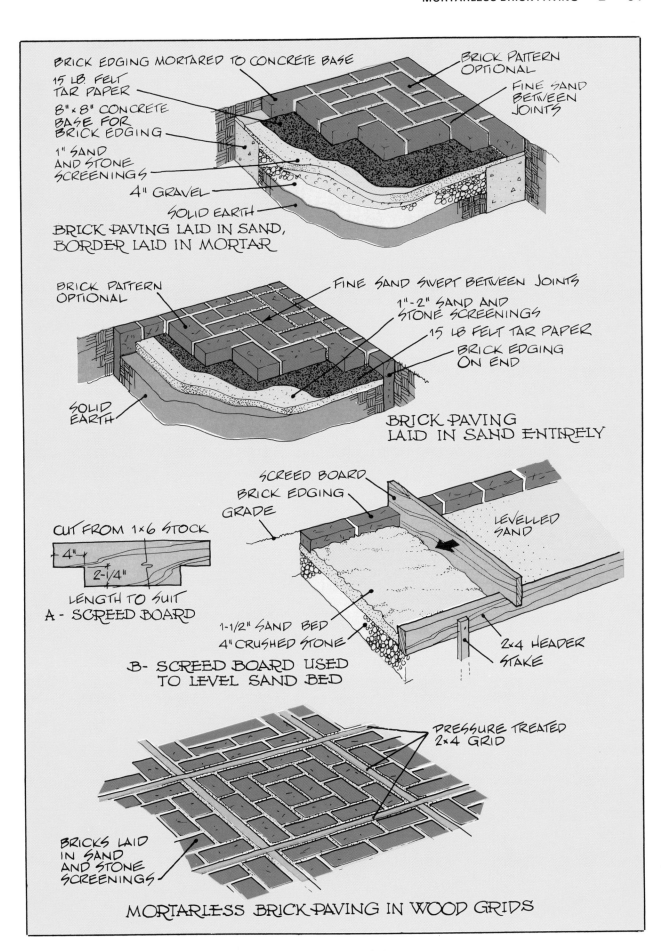

BRICK EDGING MORTARED TO CONCRETE BASE

15 LB. FELT TAR PAPER

8" x 8" CONCRETE BASE FOR BRICK EDGING

1" SAND AND STONE SCREENINGS

4" GRAVEL

SOLID EARTH

BRICK PATTERN OPTIONAL

FINE SAND BETWEEN JOINTS

BRICK PAVING LAID IN SAND, BORDER LAID IN MORTAR

BRICK PATTERN OPTIONAL

FINE SAND SWEPT BETWEEN JOINTS

1"-2" SAND AND STONE SCREENINGS

15 LB FELT TAR PAPER

BRICK EDGING ON END

SOLID EARTH

BRICK PAVING LAID IN SAND ENTIRELY

CUT FROM 1x6 STOCK

4"

2-1/4"

LENGTH TO SUIT

A - SCREED BOARD

SCREED BOARD

BRICK EDGING

GRADE

LEVELLED SAND

1-1/2" SAND BED

4" CRUSHED STONE

2x4 HEADER STAKE

B - SCREED BOARD USED TO LEVEL SAND BED

PRESSURE TREATED 2x4 GRID

BRICKS LAID IN SAND AND STONE SCREENINGS

MORTARLESS BRICK PAVING IN WOOD GRIDS

and nail a 2×4 header approximately 5' from the edge of the border for the other edge of the screed to ride on while screeding. See the illustration.

Screed about 4' at one time. Move the 2×4 header each time you want to do additional screeding. For a firm base, tamp the fill lightly before screeding. Never walk on the fill after screeding.

Lay the bricks in the bond pattern you desire, tapping each one in place with the trowel or hammer handle. A line can be attached from one end to the other when laying a long course to keep them in line. Since the bricks will settle a little, it is a good idea to lay them a bit high.

It is always more practical to work from one end to the other and make the cuts, if any, at the end. Size irregularities in the bricks are adjusted within the joints by tapping back or opening slightly.

After all bricks are laid, fill the joints between with a dry mixture of either sand or 1 part portland cement to 4 parts of sand and sweep it into the joints. The addition of the portland cement will help prevent the sand from washing out as fast.

Complete the work by dampening the top of the paving with a fine spray from a hose to settle the dry mix. Sweep in additional dry mix if needed to fill joints.

Brick patio paving or walks can be laid in squares or grids using treated 2×4s as mentioned, laying bricks inside each square. I would not recommend nailing the individual grids of 2×4s together, as it makes them too difficult to replace at a later date if they rot out. Perhaps a few finishing nails just to hold the grid together would help.

Any bond of pattern you select will work out for gridwork, but as a rule a running bond or a simple variation such as shown in the illustration is a good choice.

In my opinion, a small growth of grass between the mortar joints is not objectionable but adds to the beauty of the blocks. If this is undesirable, spray a weed killer between the blocks to keep plant growth down.

Regardless of the kind of mortarless paving selected, it is one of the more simple jobs in brickwork and provides a colorful and permanent surface with a minimum of upkeep.

HAND TOOLS

Bricklayer's trowel	Pencil
Bricklayer's hammer	Brush
Mason's rule	Knee pads or a piece of
4' level	foam rubber
Line level	Brick chisel (broad-edged)

EQUIPMENT

Wheelbarrow	Straightedge, either 2×4
Shovel and rake	or 2×6 wood 10' long
Bucket and hose	Tamper
Pick and digging iron	Broom

MATERIALS

(*Note:* No attempt is made here to estimate the amount of materials you will need for a specific job. Use the information below to make your own estimate.)

● Five bricks are used for 1 square foot. (This is half a brick more per square foot than when laying paving bricks in mortar, because of the lack of any mortar head joints.)

● Sand is sold by weight, usually by the ton or ½ ton, but you can buy any amount you need. Always ask how much the minimum delivery charge is, as you may be able to combine materials and save money.

● Stone dust is also sold by weight. It is about the same price as sand as a rule.

(*Note:* When calculating sand or stone dust, one method that works well is to figure the total square feet of the area to be covered, convert into cubic feet, and then convert into cubic yards. Using this information as a guide, it takes about 2,700 lbs. to a cubic yard or ½ ton will equal about 100 cubic feet. Adapt this to your situation or ask your building supplier to estimate, based on the above information.)

● Crushed stone or gravel may be required as a base under brick paving and will help to drain excess water away. Crushed stone or gravel is also sold by weight. The variable that works for crushed stone is 1.5 cubic yard equals 1 ton of stone. A simple formula that works for converting square feet to cubic yards is as follows:

$$\frac{\text{Width} \times \text{length} \times \text{thickness}}{27} = \text{cubic yards}$$

Remember to keep all variables in the same value such as inches or feet to obtain a correct answer. If this is a bit too complicated, your building supplier can do it for you on a calculator.

● Black felt tar paper is recommended in some cases, laid over the stone dust or sand to retard plant growth. This is optional but a good idea if the area is wooded or shaded or if there is a high degree of dampness. Estimate how much you will need by figuring the square feet of the area and then assuming a standard roll of black felt is 36" wide and will cover 400 square feet. This allows for a 1" lap over adjoining pieces.

● Mortarless brick paving can also be laid in squares between grids of wood, usually 2×4s. These 2×4s are figured by measuring the running length needed. Always buy treated lumber or a wood that will withstand dampness, such as redwood or cedar.

21
LAYING A FLAGSTONE WALK

F lagstone describes any flat stone usually not any more than 1½″ in thickness. In the eastern part of the United States, Vermont slate, New York flagstone, and natural fieldstone are used for paving, patios, walks, etc.

Vermont slate flagstone comes in different natural colors, including red, blue, green, and a dark purple. Other flagstone colors are determined by the mineral deposits found in the stone. Regardless of the type selected, ask your building dealer how he determines the square footage and price. Since stone is heavy, the delivery cost may be extra. If you have access to a rock outcropping in the mountains, there may be plenty for the collecting. Be especially careful of snakes as they love rock piles, especially in warm weather.

Flagstone is either going to be squared or irregular. The irregular shape is known as "random" and is a little cheaper to buy. Some people prefer the square to the random. It is simply a matter of choice. The square pieces, especially Vermont slate, are sawed with a straight edge in different sizes so that you have a puzzle-type arrangement when fin-

ished with no two head joints lining up. Natural flagstone picked up in the fields or mountains is almost always random and when laid in mortar presents the stone in a more rustic, natural appearance. See illustration of squared and random flagstone.

Laying Flagstone in Mortar

When laying a flagstone walk or patio in mortar, first excavate deep enough in the soil to allow about 4″ of crushed stone or gravel, a 4″ concrete slab, and the thickness of the flagstone and mortar joint. This will be about 10″ deep for flagstone of normal thickness.

Pour the concrete to the desired height, a minimum of 4″; screed off with a board; and leave untroweled on the top. Let concrete cure at least one day before working on it.

Mix the mortar as described in the materials list. If the concrete is especially dry or the weather hot, it is a good idea to dampen the concrete slab with water with a brush to help prevent the mortar from setting too fast.

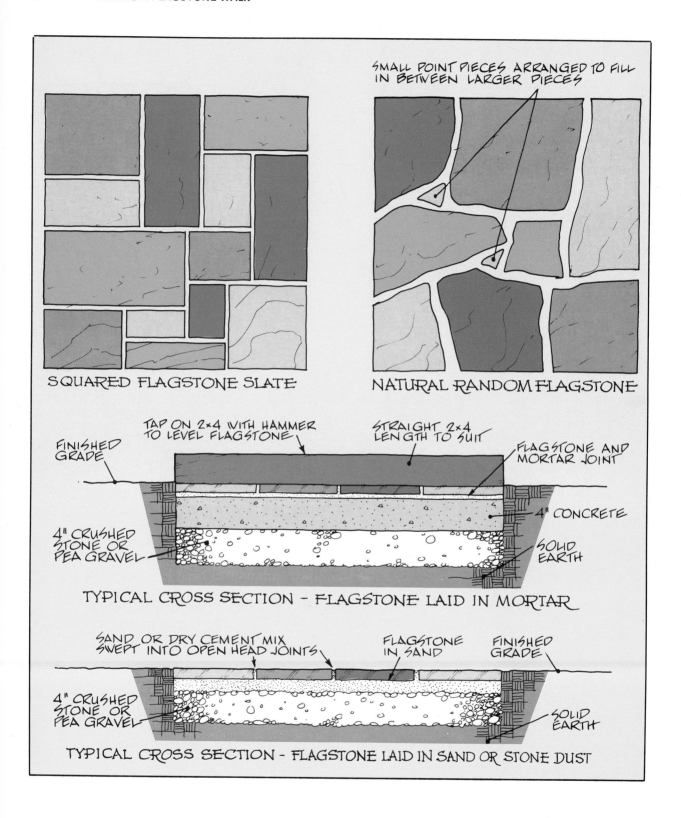

SQUARED FLAGSTONE SLATE

SMALL POINT PIECES ARRANGED TO FILL IN BETWEEN LARGER PIECES

NATURAL RANDOM FLAGSTONE

FINISHED GRADE

TAP ON 2×4 WITH HAMMER TO LEVEL FLAGSTONE

STRAIGHT 2×4 LENGTH TO SUIT

FLAGSTONE AND MORTAR JOINT

4" CONCRETE

4" CRUSHED STONE OR PEA GRAVEL

SOLID EARTH

TYPICAL CROSS SECTION – FLAGSTONE LAID IN MORTAR

SAND OR DRY CEMENT MIX SWEPT INTO OPEN HEAD JOINTS

FLAGSTONE IN SAND

FINISHED GRADE

4" CRUSHED STONE OR PEA GRAVEL

SOLID EARTH

TYPICAL CROSS SECTION – FLAGSTONE LAID IN SAND OR STONE DUST

Lay out the stones dry on the concrete for size before spreading any mortar. This gives you a chance to work the stone for best placement with a minimum of cutting. If you're laying random flagstone, you will have to fill in here and there with some small pieces of points where two angular edges meet. This is not only necessary but looks nice. However, try to hold down the number of small pieces, as they will have more tendency to work loose than large stones. See the illustration.

Mark the position of each stone laid out dry with a crayon before picking it up and spreading the mortar. Spread a coat of mortar on the concrete; do not furrow with a trowel, as it should be solid. As when paving with brick, lay the border first, leveling with a level and a 2×4. I have found that flagstone will bond better to the mortar bed if the back of each stone is plastered first and then laid in the mortar bed. This ensures that the mortar covers all bed surfaces of the stone.

If the walk is wider than 4', it is wise to attach a line slightly ahead of the stone as a gauge for level. Pick a good straight 2×4 about 5' long and lay it across the stones, settling them into position solidly by tapping down with the hammer on top of the 2×4.

I would not recommend raking out the mortar head joints and pointing them later as many people do. A much better job will be accomplished if you lay about a 5-square-foot area of stone and then point right away. The mortar will cure together better this way and is much less likely to crack later. Smooth the head joints out with the flat slicker tool or the pointing trowel. After the mortar dries enough not to smear, brush it off lightly and restrike to obtain a nice smooth finish. If the mortar starts to dry out too much in the head joints before you finish, dip the slicker tool in a can of water and restrike. This will provide enough moisture to obtain a smooth finish on the joints.

At some point it is going to be necessary to cut a stone. Cutting Vermont slate, New York flagstone, or a good flat stone can usually be accomplished by laying the edge to be cut over the edge of a straight board and scoring along that edge with the chisel. Then turn the stone over and repeat the scoring on the other side. Turn the stone back to its original position, and with a few sharp blows the stone should break neatly. Natural fieldstone should be cut along the grain whenever possible.

If just a corner or protruding area is to be cut off, it can be done by pecking along the edge with a

The combination of slate flagstone walks with an ashlar planter adds a unique charm to this home.

sharp brick hammer. Stones are unpredictable, however, and you won't win them all. If a stone breaks, either make two separate stones out of it or start over and cut another. It is always a good idea to wear safety glasses when cutting stone.

After the stonework has cured for about a week, it can be cleaned with a solution of 1 part muriatic acid to 10 parts water. Be sure to flush off the stone with plenty of clean water when finished. The cleaning will bring out the colors of the stone.

As previously mentioned, flagstone can be laid in stone screenings (stone dust) and sand as well as in mortar. This is done the same way as when laying brick paving in dry mix. Be sure that the stones are settled firmly in place by driving down with a hammer handle or a 2×4. Sweep either sand or a mixture of 1 part portland cement to 4 parts sand in between the head joints to finish the job. See illustration of elevation of stone in mortar and in stone dust.

It is, of course, more economical to lay flagstone in a dry bed than in mortar, but it is also more difficult to keep the stones in position. Regardless of which method you select, try to keep the stonework on an even plane, avoiding high and low spots in the walk. Also, don't form large head joints between the stones, as they will have a tendency to crack. Mortar head joints between flagstone should not exceed 1" for best results.

You will be down on your knees a lot when laying flagstone, and a pair of rubber knee protectors that strap on will really help. If these are not available, try a piece of foam rubber. Work a small area at a time, as stonework cannot be rushed.

HAND TOOLS

Bricklayer's trowel	Ball of line
Bricklayer's hammer	Brush
Mason's rule	Line level
4' level	Flat slicker tool
Bricklayer's broad chisel	Square and pencil
Pointing trowel	

EQUIPMENT

Wheelbarrow	Bucket and hose
Mixing box	Pick and digging iron
Shovel and mortar hoe	Tamper and broom

MATERIALS

Concrete (*Note:* See the preceding project on brick paving in mortar to estimate the cubic yards of concrete you will need.)

Reinforcement wire, to give the concrete base extra strength

Flagstone (*Note:* The amount is calculated by a specific weight equaling a given number of square feet. Different types of flagstone vary in weight according to their thickness. As a rule, however, figure that each square foot of flagstone approximately 1" thick will weigh 10 lbs. This refers to the flagstone most commonly sold by building supply dealers.)

Mortar (*Note:* Mortar for laying flagstone should be more durable and stronger than regular mortar. I recommend portland cement and lime mixed in the following proportions: 1 part portland cement to ¼ part hydrated lime to 3 parts sand. Mix the mortar stiff enough that it does not bleed or smear easily for best results. This is known as Type M mortar and is best for masonry work that comes in contact with the earth. Flagstone can also be laid mortarless the same as brick paving, with stone dust or sand as a base.)

22
STONE FIREPLACE AND BARBECUE

This project may be the answer to your vacation retreat or similar situation. Although the project looks complicated on first inspection, it really is a matter of being careful and following the plans. Have fun and enjoy the fellowship of your friends and neighbors while building the project.

Before starting the stone fireplace and barbecue, consider its location. From the practical viewpoint, it is good to locate the fireplace-barbecue so that it faces the direction of the prevailing wind. Although winds do not always blow from the same direction, in most localities they blow from one direction at least half of the time; therefore, proper attention to this detail will result, probably, in the cook's not getting smoke in his face more than half of the time. The fireplace should not be located where a building or tall tree is likely to produce down drafts of air nor where the smoke and hot gases from its flue will damage shrubbery or trees. A mistake that is sometimes made is to build a fireplace under a tall tree on the assumption that, since the tree's branches are located 20' or more above the flue, no damage will result. However, damage can be done,

and done quickly, if wind and weather conditions are such that the flue gases ascend vertically for even a few minutes without much mixing with the atmosphere.

In some localities there are ordinances that regulate the spacing of such structures as fireplaces in relation to property lines. Most building codes now provide that dwelling houses must not be placed closer than a specified number of feet from the boundaries of the property, and sometimes these same provisions apply to garages, fireplaces, and any other type of structure that projects above the ground. Application of these laws to outdoor fireplaces is the exception rather than the rule, but, if you live in a community where building is strictly regulated by a code, it is better to find out in advance than to be required to tear down the stonework after it has been constructed.

Privacy and accessibility are other important factors in choosing the location. An outdoor fire carries with it an age-old implication of sharing the experience with others. Nevertheless, a certain amount of privacy is desirable. Usually the fire-

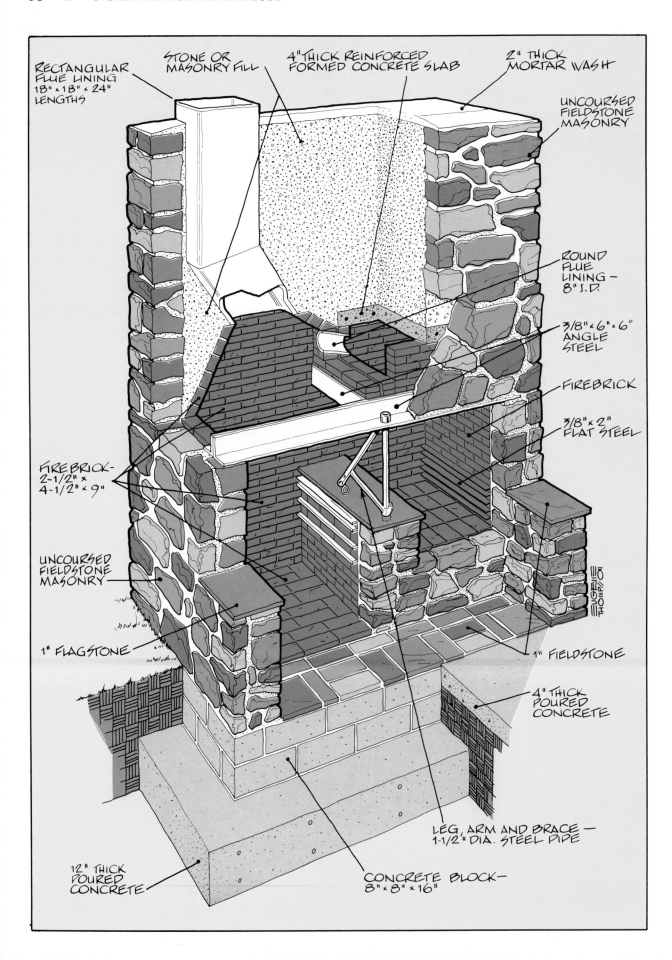

RECTANGULAR FLUE LINING 18" x 18" x 24" LENGTHS

STONE OR MASONRY FILL

4" THICK REINFORCED FORMED CONCRETE SLAB

2" THICK MORTAR WASH

UNCOURSED FIELDSTONE MASONRY

ROUND FLUE LINING – 8" I.D.

3/8" x 6" x 6" ANGLE STEEL

FIREBRICK

3/8" x 2" FLAT STEEL

FIREBRICK – 2-1/2" x 4-1/2" x 9"

UNCOURSED FIELDSTONE MASONRY

1" FLAGSTONE

1" FIELDSTONE

4" THICK POURED CONCRETE

LEG, ARM AND BRACE – 1-1/2" DIA. STEEL PIPE

12" THICK POURED CONCRETE

CONCRETE BLOCK – 8" x 8" x 16"

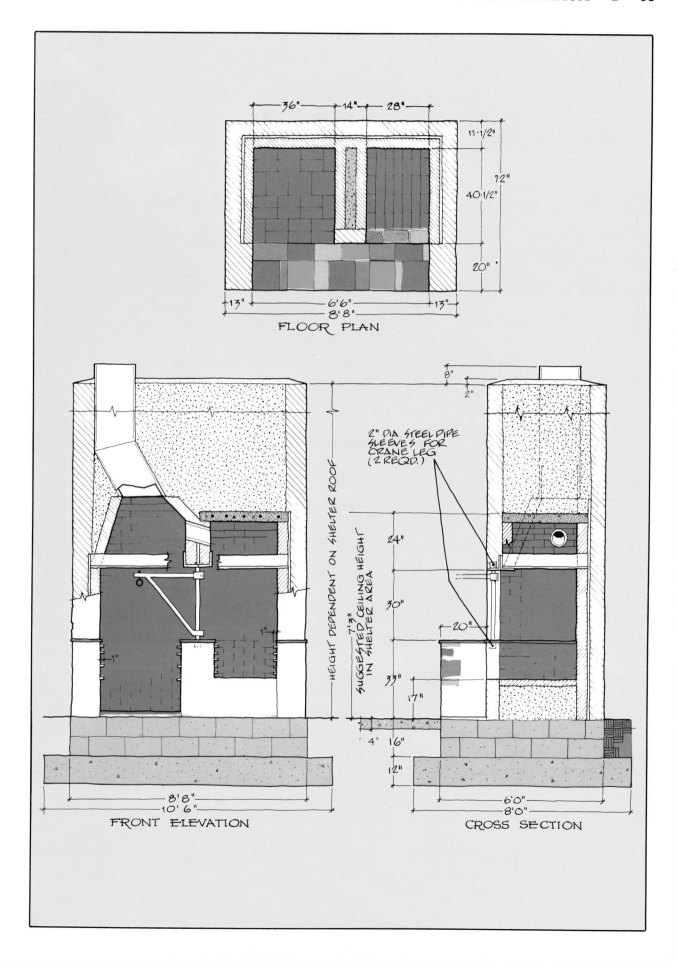

FLOOR PLAN

FRONT ELEVATION

CROSS SECTION

place is located so that it cannot be readily viewed from the public street. The same attention, however, is not paid to screening it so that it cannot be seen from the back yards of adjacent properties as is paid to the screening of a terrace used for sunbathing. However, personal taste should rule here. Accessibility to the kitchen in the dwelling or to some other building where pans, charcoal, etc., can be stored is a convenience but not an absolute requirement.

Once the location has been decided upon, start the construction of the stone fireplace and barbecue by staking off with a line where the footings go. Excavate below the existing freeze line for your area. The amount of concrete for this footing is obviously more than one would want to mix with a hoe. Therefore I would recommend ordering a 5-bags-of-portland-cement-to-the-cubic-yard mix from a concrete company. This is also known as a prescription mix of 2,500 psi in many areas.

Lay the two courses of block up to finished grade line as shown on the plan. Use the same mortar mix as for the stonework. Fill inside the foundations solid with block work.

Lay the stone walls out as shown on the plans. Firebricks were used to line the inside of the fireplace and cooking area, but only because the builder happened to have a large number of these on hand. They are more expensive than regular bricks. A good, hard red brick can be substituted and is perfectly acceptable.

As the firebrick walls are built, build in the barbecue supports of flat steel in the side walls, letting them project 1″. This spacing of the steel supports allows the grate to be put at the desired cooking height.

The center partition wall is laid up to a total height of 33″ as shown in the cross-sectional view and finished off with flagstone top.

A 2″-diameter steel pipe sleeve is built in the center of the pier approximately 2″ back from the face of the wall. This is for the crane that will support the kettle and chain. The triangular crane can be made from some lengths of 1½″ pipe with two stout metal clamps at the bottom and top edges. Where the 45° piece meets the 90° piece, it can be welded together for strength. When built into the top angle iron the same way, it makes a swinging crane that will more than hold the iron kettle filled with water. See L on front elevation.

As the stonework is laid and the mortar joints stiffen, they are raked out slightly with a piece of rounded-off broom handle and brushed to close up any holes and remove excess particles of mortar.

The longest angle irons, which cover the front opening, should rest on each side of the stonework about 8″ for support and are placed in a back-to-back position forming a lintel over the front. See G on front elevation and cross section.

A hole is burned with a cutting torch through the angle iron at the point over the center partition where the top sleeve goes to support the top of the crane. This allows the center pipe of the crane to be slipped in place. The arm is attached by heavy-duty clamps. See front elevation.

The angle irons that support the center partition wall at the top of the firebox are laid on the long front irons and built into the back walls. Firebricks are laid on the irons. See G on cross section and front section.

A piece of round flue lining with an inside diameter of 8″ is built through the approximate center of the partition wall to serve as a flue to carry smoke from the right side of the barbecue to the left side, which contains the flue lining that forms the chimney. See front elevation H and C.

A concrete reinforced slab is formed and poured over the top of the right side of the barbecue, sealing it off and making it airtight. See detail I on front and cross-sectional views.

The firebrick walls at the point where the flue lining is to set are corbeled out for the 18″×18″×24″ flue forming the chimney. All masonry work around the flue is filled with solid masonry fill. The chimney is built from this point to the top and filled in solid with masonry fill.

At the top of the chimney, the flue lining extends about 6″ above the stonework. A mortar wash coat is applied to seal off the top and help air currents flow smoothly. See front-elevation view.

The outer hearth of flagstone is laid last, as shown on the cross-sectional view and the photograph. Vermont slate was used for this because of the pleasing color of the stone.

The entire chimney and fireplace should be cleaned down with a solution of 1 part muriatic acid to 10 parts water after it has cured for about a week. Rinse the work with a hose with a nozzle to bring out the full color of the masonry work.

This project was designed and built by a group of retired men at a development called Glenn Haven in West Virginia as a central gathering place and receives hard use throughout the spring, summer, and fall of the year. None of these men were masons by trade, but they had the time and imagination to create this useful, beautiful fireplace and barbecue. It has seen many a corn, chicken, and steak cookout, and it is a testimonial to its builder's craftsmanship and initiative.

HAND TOOLS

Bricklayer's trowel	Flat slicker jointer
Bricklayer's hammer	Rounded end of a broom
2-lb. hammer	handle, about 5" long
Assortment of chisels:	Ball of line
broad-edged, point,	Line pin and set of line
and regular mason's	blocks
cutting chisel	25' steel measuring tape
4' level	Mason's pocket rule
2' level	Square
Pointing trowel	Brush

EQUIPMENT

Wheelbarrow	Pick and digging iron
Mixing box	Sledgehammer
Shovel and mortar hoe	Scaffolding

MATERIALS

For Footings

3 cubic yards concrete
Concrete reinforcement wire, about 10'6"×8'

For Foundation Walls

111 concrete blocks, 8"×8"×16"
2 bags Type 1 portland cement
2 bags mason's hydrated lime
¼ ton sand

For Walls and Fireplace above Grade

Approx. 436 firebricks, 2½"×4½"×9", for lining the interior firebox and cooking areas
Approx. 30 square feet flagstone, for outer hearth and top of walls
1 length round flue lining, approx. 14" long and 8" inside diameter, for through-wall flue above barbecue area to carry smoke to main flue
Rectangular flue linings, 18"×18"×24" (*Note:* Number needed will depend on height of chimney.)
2 angle irons, 6"×6"×⅜"×7'6"
2 angle irons, 6"×6"×⅜"×40"

14 pieces flat steel, 2"×½"×⅜"×40", for adjustable barbecue-rack supports
1 steel pipe arm, 1½" in diameter, and brace with sleeves, for swinging arm of pot crane
1 wheelbarrow concrete and 1 piece reinforcement wire, for forming small reinforced concrete slab over right-hand side of barbecue area
Barbecue grates (*Note:* Have made by welder a grid of ½" rebar. Allow ¼" on each side for clearance when measuring for size.)
1 drip pan, heavy-duty galvanized metal supported by a ½" steel rebar frame welded together (*Note:* Chain and iron cooking kettle were purchased at a farm sale.)
Rubblestone (*Note:* This is estimated by the cubic foot as a rule. The average rubblestone weighs about 125 pounds per cubic foot. To determine how much stone you will need for your project, place on a freight scale a stone that measures approximately 12"×12". Record this weight and when buying the stone, check the total weight bought and divide by the weight per cubic foot. Stones are usually sold by the ton or ½ ton. If you can find an old stone fencerow, a rock outcropping in the mountains, or a collapsed barn or stone house, then your only cost will be carting stones home. Remember, however, to get permission first!)
Mortar (*Note:* In my opinion the best mortar for stonework is always a portland-cement base mortar, as it is stronger and adheres to the stone better than any other type. Use 1 part portland cement to 1 part hydrated lime to 6 parts sand, mixed with water to a stiffness that will support the weight of the stone. The quantity needed will vary with the size and shape of the stone and the thickness of the walls, but an average is 8 cubic feet of mortar per cubic yard of stonework. If you prefer, buy two sacks of portland cement, two sacks of hydrated lime, and a ton of sand. Use this and see how far it goes. Using this as a guide, you should be able to estimate pretty closely how much you need to finish.)

23
BRICK GATE WALL
AND ENTRANCEWAY

This project will really dress up the entrance to your yard and provide an attractive method of closing it off. The suggested size of the gate opening is considered to be standard; however, you may make it larger if needed. No footing is poured under the gate itself, as it is not needed.

Start out by staking off the area with a line where the footing is to be poured. Excavate to the frost line for your area. The plan shows 22", which is acceptable in the Middle Atlantic area where I live.

Mix and pour the concrete, using 1 part portland cement to 2 parts sand to 4 parts crushed stone or gravel, mixed with water.

Lay two courses of 12" concrete blocks up to grade, making sure that they are in line with each other on both sides of the gate opening. Mix the masonry cement mortar, 1 part masonry cement to 3 parts sand or ½ bag to 8 shovels sand.

Lay out the first course of the brick piers and walls for both sides of the entranceway, using a line stretched across the entire project to make sure that all the work is in line. The best procedure to follow is to dry-bond the bricks and then lay them

mortar forming about a ⅜" mortar head joint between each brick. This way you will not lose the brick spacing.

As the brickwork progresses, it can be done most efficiently by building the piers up on each end, then attaching a line to the end piers and building the piers between. Lay the brick walls between to another line as the piers are built, as there is a 2" setback from the face of the piers on both sides of the wall (see top section view of wall). This setback provides a distinct break or depth appearance and highlights the wall from the piers.

Tool the mortar joints when thumbprint-hard with a convex (round) jointer or one of your choice. Point up the flat mortar joints neatly where the piers and wall meet in the angles.

Build in wall ties every three courses to tie the piers to the wall. Lay brick courses for height to number 6 on the modular rule.

Build the piers ahead of the wall at the point where the curved section starts to the top. Project the rowlock brick shown on front elevation out all sides of pier about ¾". Use solid bricks for rowlock

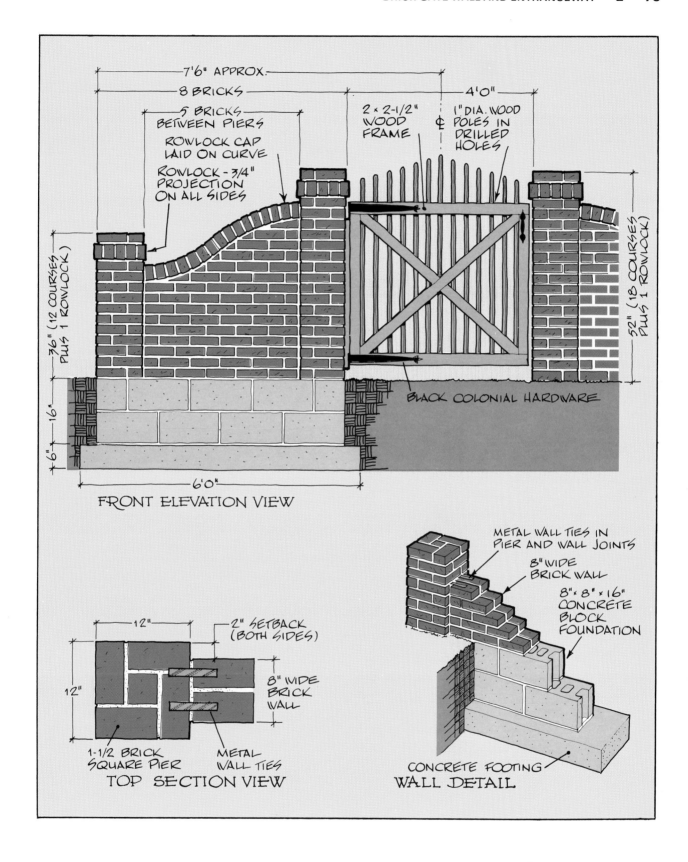

7'6" APPROX.

8 BRICKS

5 BRICKS BETWEEN PIERS

ROWLOCK CAP LAID ON CURVE

ROWLOCK – 3/4" PROJECTION ON ALL SIDES

2 x 2-1/2" WOOD FRAME

1" DIA. WOOD POLES IN DRILLED HOLES

4'0"

36" (12 COURSES PLUS 1 ROWLOCK)

52" (18 COURSES PLUS 1 ROWLOCK)

16"

6"

6'0"

BLACK COLONIAL HARDWARE

FRONT ELEVATION VIEW

12"

2" SETBACK (BOTH SIDES)

8" WIDE BRICK WALL

12"

1-1/2 BRICK SQUARE PIER

METAL WALL TIES

TOP SECTION VIEW

METAL WALL TIES IN PIER AND WALL JOINTS

8" WIDE BRICK WALL

8" x 8" x 16" CONCRETE BLOCK FOUNDATION

CONCRETE FOOTING

WALL DETAIL

so there is no bother having to fill up holes later on the sides. Your local brick supplier will stock solid bricks for this purpose to match almost any blend.

To make a pattern for the curved portion of the wall between the piers, cut a piece of plywood to the approximate curve shown on front elevation. Then as the brick courses are laid, hold the pattern against the wall, mark the bricks to the curvature, and cut to fit with a chisel and hammer. It will take you a little time to do this, but it is not that hard to cut.

Be especially accurate with the cutting and laying of the bricks on the curvature, as the mortar joints under the rowlock cap course should be the same thickness to look right.

Tool the mortar joints on top of the wall flat with the slicker tool, making sure that there are no holes for water to penetrate. Brush when dry enough not to smear.

After the brickwork is completed and has cured for about a week, wash it down with a solution of 1 part muriatic acid to 10 parts water, flushing the work well with water when finished to remove all dirt and excess chemical cleaner.

The gate is built of 2"×2½" wood frame using 1" closet poles drilled and inserted as shown on the plan. Attach the hinges to the gate and to the wall by drilling and inserting lead anchors (sleeves) in the brickwork. A good stout colonial strap hinge makes a nice appearance and will support the gate well. All materials for the gate are available from your local building supply or hardware store. Closet poles shown as a part of the gate could be replaced with pickets if you like, which would make the gate easier to build.

A nice flagstone or brick walk can be built later to finish the job.

HAND TOOLS

Bricklayer's trowel	Brick chisel (wide-bladed)
Bricklayer's hammer	Ball of line
25' measuring tape	Flat slicker jointer
Mason's modular rule	Chalk line
4' level	Convex jointer
2' level	Brush

EQUIPMENT

Mixing box	Bucket and hose
Wheelbarrow	Mortarboard
Shovel and mortar hoe	Pick and digging iron

(*Note:* In addition, some basic carpentry tools such as a square, handsaw, and clawhammer will be needed to build the gate and cut out a wood template form for the curved part of the wall.)

MATERIALS
For Footing

Approx. ¼ cubic yard concrete (2 bags portland cement, 400 lbs. sand, 450 lbs. crushed stone or gravel)

For Walls

16 concrete blocks, 12"×8"×16"
625 standard bricks, for both sides of the project
6 bags regular masonry cement
¾ ton sand
24 metal corrugated masonry wall ties, to tie the brick piers to the walls between the piers.

(*Note:* Lumber, closet poles of wood, and hardware are not shown. Your local lumber supplier or building supply store can help with this according to the material in stock. The same is true for the hardware on the gate. The gate is fairly simple to build but can be substituted with another type if you so desire.)

24
ROUND BRICK BULL'S EYE OPENING

The circular brick opening shown in this project is a real eye-catcher and relatively simple to build. It looks great in a garden wall to break the monotony of a straight wall and will enhance any patio wall. No frame or glass was used in the opening when it was completed, as the builder wanted the effect of the open circular window. This type of circular opening is known as a "bull's-eye" in the construction trade and consists of a completely circular arch.

The brick wall is laid in running bond with a half lap. No brick headers were used; therefore metal wall ties were needed to tie the wall together. The wall ties were laid across the two widths of wall every six courses in height or approximately 16". These corrugated wall ties are readily available from a local building supply dealer for a few cents each.

The bricks are laid on a wood form for support, then the form is removed after the wall has hardened.

Start by excavating the footing for the wall below the frost level in your locality. Mix and pour a footing of concrete in the proportion of 1 part portland cement to 2 parts sand to 4 parts crushed stone. Concrete block can be used below the finished grade level for the foundation.

If you use old bricks, as in the photo, don't be concerned if they do not work perfectly for bond and you have to cut one here and there, as used brick will vary considerably. You can obtain these old bricks where an old building is being torn down or around an urban redevelopment project. Old barns, sheds, etc., are all sources of used bricks. Check your local newspaper, as many times they will be advertised there. However, make sure the bricks are solid and not soft, as soft bricks will not last long once laid in a wall and exposed to the weather before flaking off. Take a hammer along with you; if a brick emits a good solid sound when struck lightly with the hammer, it should be okay to use.

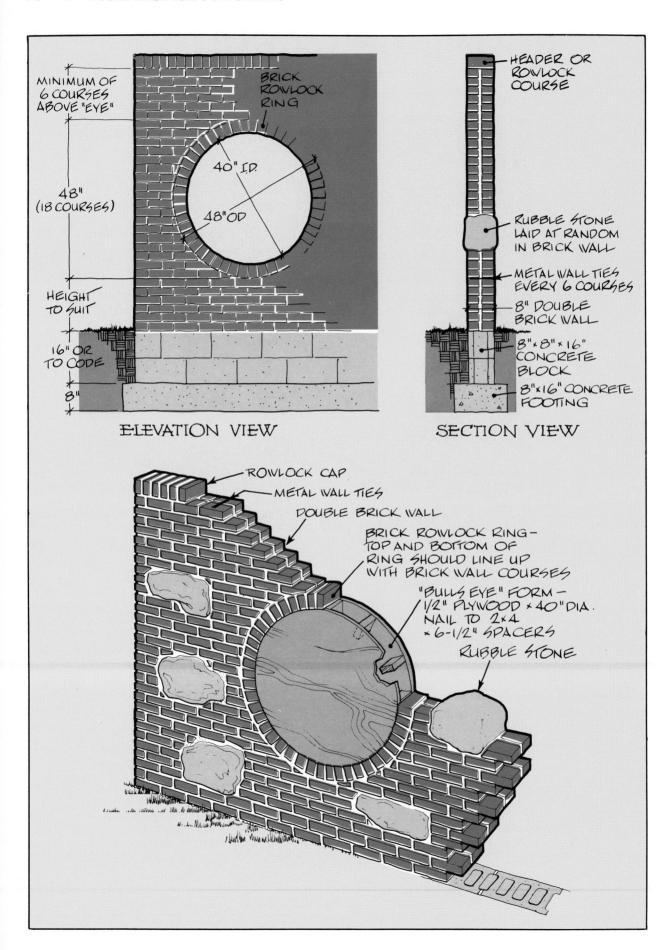

MINIMUM OF 6 COURSES ABOVE "EYE"

BRICK ROWLOCK RING

40" I.D.

48" OD

48" (18 COURSES)

HEIGHT TO SUIT

16" OR TO CODE

8"

ELEVATION VIEW

HEADER OR ROWLOCK COURSE

RUBBLE STONE LAID AT RANDOM IN BRICK WALL

METAL WALL TIES EVERY 6 COURSES

8" DOUBLE BRICK WALL

8" × 8" × 16" CONCRETE BLOCK

8" × 16" CONCRETE FOOTING

SECTION VIEW

ROWLOCK CAP

METAL WALL TIES

DOUBLE BRICK WALL

BRICK ROWLOCK RING — TOP AND BOTTOM OF RING SHOULD LINE UP WITH BRICK WALL COURSES

"BULLS EYE" FORM — 1/2" PLYWOOD × 40" DIA. NAIL TO 2×4 × 6-1/2" SPACERS

RUBBLE STONE

It is a good idea to dampen old bricks slightly with water and brush before laying them in the wall to reduce premature setting in the mortar. This is especially true if the bricks are not very hard or if the work is being done in hot dry weather.

Lay the brick courses to a height of number 6 on the mason's course counter spacing rule. Tool the mortar joints as soon as they are thumbprint-hard or they will have a tendency to turn black as the metal jointer is run through them. Lay wall ties across the brick wall every six courses to bond it together. Lay the stone randomly according to your preference as the wall is built. Don't overdo it, however, as too many stones will make it look cluttered.

Study the plan and photograph for an idea of what the completed project should look like. When the wall is built up to the bottom of the circular opening or where it starts, set the arch form on the wall and brace it in place. This form can be built of two pieces of ½" plywood with 2×4 cross bracing nailed on the inside. The brick rowlock spaced around the circle is marked on the outside edge of the form using the mason's course counter rule. Make it work in full bricks with no cuts.

After the form is built and braced into position, lay the brick wall to the line, building in the arch bricks on the form as the wall goes up. This keeps all of the work in a straight line and tied together well. See detail on plan of how arch form is constructed. Note that the arch form is 40" in diameter and 8" wide to carry the brick rowlock. Space the brick courses for height as shown. It will take 11 courses, and the last brick laid should be even with the top of the circular ring of the arch.

Continue building the wall to the height desired with at least six courses laid over the top of the opening. The last course laid on the wall should be a solid brick cap to prevent water from entering. A header or rowlock course will finish the job.

Wait at least two days after the project has been completed, then carefully remove the arch form. Neatly trim out any hard mortar in the joints with a chisel and repoint with fresh mortar.

Be especially careful not to chip the edges of the bricks as this is done. Have no fear—the circular opening will support the brickwork laid over it better than any lintel could.

Let the mortar cure for about a week and if needed clean with a solution of 1 part muriatic acid to 10 parts water. Be sure to wet the wall first and rinse it well afterward.

This project provides the handyman with a lot of liberty in designing a wall utilizing the round brick circular opening. You may decide to build a couple of these in one wall, rather than one, for a really unusual appearance. This should be one of the most enjoyable and creative projects shown in the book. Have fun!

HAND TOOLS

Bricklayer's trowel	Ball of line
Bricklayer's hammer	Set of line blocks
Mason's course counter spacing rule	Convex jointer
	Slicker jointer
4' level	Pointing trowel
2' level	Brush
Brick chisel (wide-bladed)	

EQUIPMENT

Mixing box	Bucket and hose
Wheelbarrow	Mortarboard
Shovel and hoe	

MATERIALS

Concrete (*Note:* The amount of concrete needed for the footing will depend on the length of the wall.)

Brick (*Note:* Estimate seven bricks per square foot of single wall thickness. Since this project is a double wall (8" thick), be sure to double your estimate to take care of both sides.

Masonry cement (*Note:* Each bag will lay 125 bricks.)

Sand (*Note:* Estimate 1 ton of sand per 1,000 bricks. Since the stones spotted (laid) throughout the wall are few, no extra allowance is made for mortar to lay them.)

Stones (*Note:* Random stones selected from a rock outcropping or an old stone fence will work fine for the few you will need. If you do not have access to these, a masonry supply dealer should stock some suitable type of stone.)

25

RUBBLESTONE RETAINING WALL

A stone retaining wall not only serves a useful purpose in holding back the earth, but can be used as a flower or planting bed to improve the general appearance of your home by adding a natural rustic look. The retaining wall shown in the photograph did not need any metal reinforcement because of its battered design.

Start by staking off a line where the wall is to be built. As usual, be sure to excavate below the freeze line for your area. Notice on the section view of the plan that an extra-wide footing is shown. This is done because the wall is wider at the bottom than at the top, so that it will better resist any pressure from the backfill. The back of the wall line is built on an angle to the top and is called a "battered wall."

Mix the concrete to a proportion of 1 part portland cement to 2 parts sand to 4 parts crushed stone with water.

After the footing has cured for at least a day, lay the block foundation, or stone if you have plenty, up to the grade line as shown on the plan.

The rubblestone wall shown in the photograph was built next to a blacktop driveway. Be sure to start your stonework so that no concrete block shows above the blacktop.

Build in the drainpipe through the wall approximately every 8' to provide proper drainage and relieve any pressure that may build up. In addition, the flexible drainpipe should be installed with a slope away from the wall so the trapped water will drain away. Install the drain tile through the wall on a little slope also so the water will drain easily.

Tool the finish of the stonework by raking out the mortar joints about ½" to ¾" and then smoothing with the flat slicker tool or the pointing trowel for larger joints. Don't attempt to rake out the joints until they have set enough that the stone will not settle out of position.

Cut the stones where needed to fit and fill in fully with mortar around all edges. Continue drawing the stonework in the back until the top of the wall is reached. The completed width at the top should be about 12".

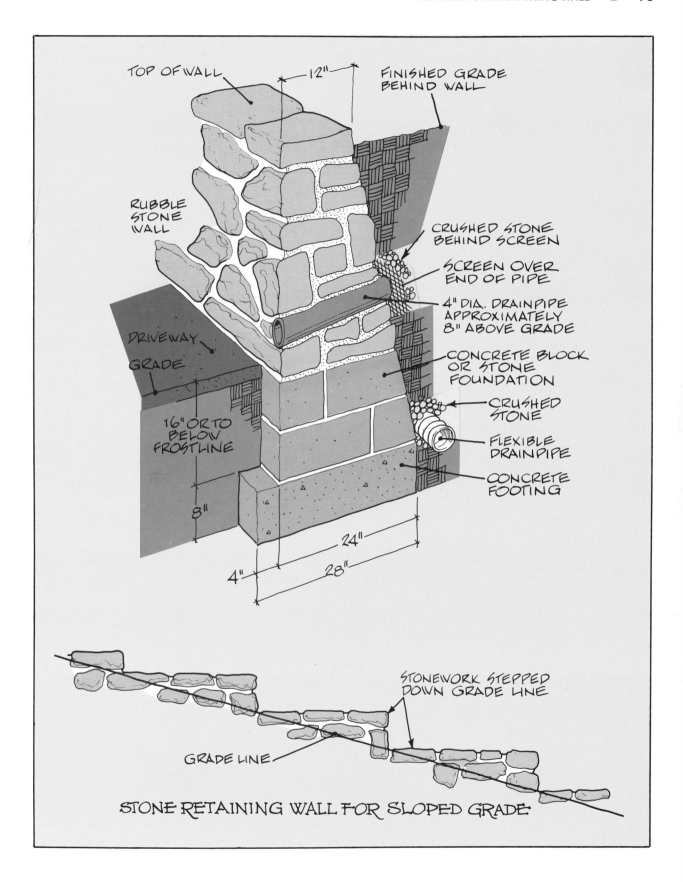

TOP OF WALL

12"

FINISHED GRADE BEHIND WALL

RUBBLE STONE WALL

CRUSHED STONE BEHIND SCREEN

SCREEN OVER END OF PIPE

4" DIA. DRAINPIPE APPROXIMATELY 8" ABOVE GRADE

DRIVEWAY GRADE

CONCRETE BLOCK OR STONE FOUNDATION

CRUSHED STONE

16" OR TO BELOW FROSTLINE

FLEXIBLE DRAINPIPE

CONCRETE FOOTING

8"

24"

4"

28"

STONEWORK STEPPED DOWN GRADE LINE

GRADE LINE

STONE RETAINING WALL FOR SLOPED GRADE

After the mortar has dried enough not to smear, brush it with a soft brush to remove any particles of mortar. This should also remove most of the dirt particles on the wall. As a rule it is not necessary to wash stonework with a chemical cleaner, but if you want to bring out the color try a solution of 1 part muriatic acid to 10 parts water. Rinse the work thoroughly after cleaning to remove all chemicals.

After the wall has cured for about a week, fill behind with earth and level off with a rake.

A retaining wall will retain earth on a slope of falling grade very efficiently if built correctly. Just make sure that a line is stretched along the slope and the stonework is stepped above the grade. See illustration of stonework stepped on falling grade.

HAND TOOLS

Bricklayer's trowel	Pointing trowel
Bricklayer's hammer	4' level
50' measuring tape	Brush
Mason's pocket rule	Heavy hammer (2-lb.
Ball of line	suggested)
Brick chisel (wide-bladed)	Wide-bladed slicker tool

EQUIPMENT

Mixing box	Bucket and hose
Wheelbarrow	Digging iron and rake
Shovel and hoe	Mortarboard

MATERIALS

(*Note:* The amount of materials needed will depend on how long and deep in the ground your wall will be. Listed here are the variables needed to figure materials for the job.)

● Concrete is estimated by the cubic yard, with 27 cubic feet to a cubic yard.

● Rubblestone is readily available, usually at no or little cost in the mountains, from old stone fences and barns that are in a state of collapse, etc. Scout around and there should be no problems in locating these for the asking and a little work on your part. However, always be sure to ask before removing any stone from someone else's property.

● Mortar should be mixed to a proportion of 1 part portland cement to 1 part hydrated mason's lime to 6 parts sand. The best method I know to determine how much mortar is needed is to buy a couple of bags of portland cement and hydrated lime and see how many square feet it does for the wall you are building. Then, use this information as a base for the rest of the job. Mortar for a big stone job is figured by the cubic yard, but this would be entirely too much for a small job.

● A supply of concrete or tile drainpipe, 4" in diameter, is needed for through-the-wall drains as shown on the plan. Some crushed stone is needed for around the back of each drainpipe and under and over the flexible drainpipe at base of footing. Flexible 4" drainpipe is available in rolls from your building supply dealer.

● A nonrusting piece of screen should be placed over and around the back of each drainpipe where it passes through the wall to prevent mud from clogging it.

26
BUILDING BRICK STEPS AND PORCH

This set of steps has four risers and three treads for a total height of 28″ and step width of 36″. Study the plan to see how this is worked out. You can enlarge the number of steps and the length by using the basic measurements adjusted for your specific needs. The riser is the vertical part of the step, the tread is where one places the foot when climbing, and the top part of the porch is known as the platform.

Start by staking off with a line where the footings are to be excavated. Use the plan as a guide, working out the length of the steps according to your requirements.

Mix and pour the concrete below the existing frost line to a proportion of 1 part portland cement to 2 parts sand to 4 parts crushed stone.

After the footing has cured for a day, lay the blockwork up to grade line as shown on plan. A standard depth of 24″ is shown on the plan for the area where I live, since this would be below the frost line.

Assuming that the steps you build will be the same as shown on the plan, build up the porch first as shown. This is 28″ high and 40″ wide. Nine courses of bricks are used, with a rowlock top as shown. The rowlock is projected out about ⅝″ to form a water drip and for appearance. Slope the top of the porch platform from the back to the front about ¾″ to allow water to drain off. Use a line on all brickwork over 4′ in length for best results.

The steps are marked off in divisions of 12″ in depth and 7″ in height, allowing a slope from front to back of about ¼″ to allow for water drainage. Project the front edge of each step ⅝″ to form a drip edge. Build the rowlock header back against the porch wall as shown on the plan.

The steps consist of a course of stretcher bricks with a rowlock on top. I would suggest laying about three bricks on each end of the step, attaching a line, and filling in between. Each step would appear as in the front view shown on plan.

If you are accurate and build each step to the correct height and width as shown on the plan, they will work out evenly. Notice on the side elevation that each step is exactly 6¾″ at the back and 7″ at the front. This is due to the ¼″ slope mentioned earlier to drain water off. This is important to keep ice from forming on the step in winter.

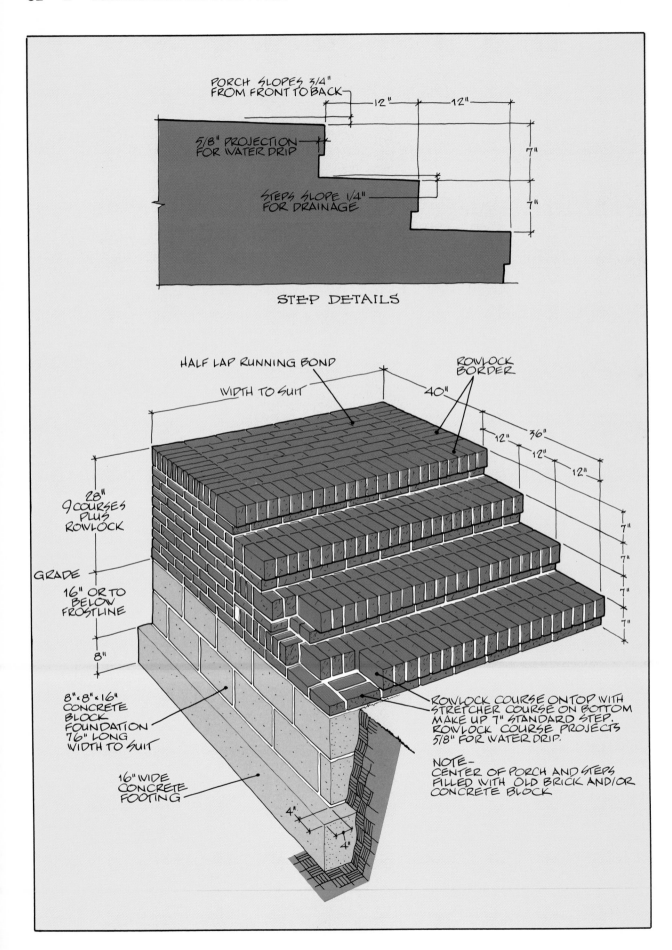

PORCH SLOPES 3/4"
FROM FRONT TO BACK

12"   12"

5/8" PROJECTION
FOR WATER DRIP

7"

STEPS SLOPE 1/4"
FOR DRAINAGE

7"

STEP DETAILS

HALF LAP RUNNING BOND

ROWLOCK
BORDER

WIDTH TO SUIT

40"

12"   36"

12"

12"

28"
9 COURSES
PLUS
ROWLOCK

7"

7"

GRADE

7"

16" OR TO
BELOW
FROSTLINE

7"

8"

8"×8"×16"
CONCRETE
BLOCK
FOUNDATION
76" LONG
WIDTH TO SUIT

ROWLOCK COURSE ON TOP WITH
STRETCHER COURSE ON BOTTOM
MAKE UP 7" STANDARD STEP.
ROWLOCK COURSE PROJECTS
5/8" FOR WATER DRIP.

16" WIDE
CONCRETE
FOOTING

NOTE—
CENTER OF PORCH AND STEPS
FILLED WITH OLD BRICK AND/OR
CONCRETE BLOCK

4"

4"

Fill in the center of the steps and porch with old used bricks or concrete blocks as the work progresses.

As the mortar joints become thumbprint-hard, tool them with the convex jointer, except on the top of the treads and platform, where I would recommend using the flat slicker tool. Brush the joints as soon as they have dried enough not to smear. Use solid bricks with no holes in them for the edge of the steps as shown on plan.

The top of the porch platform is shown on the plan with a brick rowlock border and brick paving laid in running bond in between. The variation of brick positions between the border and the running bond make a stronger top platform. Use the mason's scale rule to work out the brick rowlock cap evenly with no cutting, the same as in the steps.

Let the project cure for about a week, then wash it down with a solution of 1 part muriatic acid to 10 parts water, rinsing off well with water to remove all chemicals when finished.

The advantage of a brick porch and steps is not only their beauty, but no maintenance over the years. The most important thing to watch as you build this project is to keep the porch to the specified height and the risers and treads the correct size so that each step will work out even and there will be no difference that could cause a person to stumble when using them. If a minor difference does occur because of varying lengths of bricks, it can usually be compensated for in the mortar joints. It really doesn't matter if each step is 12½" deep rather than 12" as long as all the steps are the same.

I consider building steps one of the more challenging projects in bricklaying. Don't rush the job. The finished job will be well worth the efforts. Wrought-iron railings could be attached after the brickwork is completed.

HAND TOOLS

Bricklayer's trowel	2' square
Bricklayer's hammer	Ball of line
Mason's rule	Brush
4' level	Convex jointer (round)
2' level	Flat slicker tool
Bricklayer's broad chisel	Pair of line blocks
Pointing trowel	

EQUIPMENT

Wheelbarrow	Shovel and mortar hoe
Mixing box	Bucket and hose
Mortarboard	Pick and digging iron

MATERIALS

(*Note:* No specific amounts of materials are given for this project; you will have to do your own estimating based on the information below. The height and width, however, are stated in order to explain how the steps are worked out to use full bricks and have a 7" riser.)

Concrete, for footing under porch walls and under area where steps are to be built (*Note:* Amount will depend on size of the porch and steps you decide to build.)

Reinforcement wire, to be placed in concrete where steps are to be built

Bricks (*Note:* Figure seven bricks per square foot of wall area. If the bricks you use have holes in them, called coring, be sure to buy about a dozen solid ones for the edge of the brick steps. Almost all brick suppliers stock solid bricks to match for this purpose.)

Concrete block, for foundation equal three blocks to every 4' in length and one course to every 8" in height. Inside area of porch should not be excavated, only where the walls are to be built. If you need a number of blocks to fill in under the steps and porch, check with your local block plant for seconds, which will be about half the price and perfectly acceptable.)

Mortar (*Note:* Allow 125 bricks or 25 blocks per bag of masonry cement.)

Sand (*Note:* Estimate 1 ton sand per 1,000 bricks.)

27

OUTDOOR BRICK BARBECUE AND CHIMNEY

This is a most unusual barbecue and chimney and utilizes arches for extra beauty. The project is built around the metal fireplace grille unit. There is no flue lining used inside the chimney, as not enough heat passes through it to require lining. The handyman who built this project is a professor of music at a college and not a professional mason. This project is a good example of what a handyman can accomplish if he takes his time and has a little imagination.

This barbecue was built at the corner of a patio with arched openings next to it. You may or may not decide to build the arched openings.

Start by staking out with lines where the footings are to be dug. Excavate to a depth exceeding the existing frost line for your area. Mix and pour the concrete footing in the proportions of 1 part portland cement to 2 parts sand to 4 parts crushed stone, with water. It takes only ¾ cubic yard for the footings, and it would be very expensive to order the amount from a concrete plant.

Study the layout plan and lay out the 8″ and 4″ block walls as shown. Leave the area in the center of the barbecue hollow, and do the same for the brick storage cabinets. You should have the metal fireplace unit and Dutch doors before starting on the brickwork to make sure they will fit.

As the brickwork proceeds, insert metal wall ties about every six courses of brick to tie the angles together. Lay all brickwork to number 6 on the modular rule. This will be three courses of bricks to every 8″ in height. Drop the storage boxes off when the height shown on the plan is reached. They will have flagstone tops that project out 1″ on all sides.

The metal Dutch doors should be built into the brick storage cabinets when needed. The same is true for the small cast-iron door centered under the metal grille unit. Leave the correct-size opening for the metal grille unit and build the masonry up to the point where the arch starts or the top of the grille. Build a small wooden arch form to support the brick arch over the grille opening.

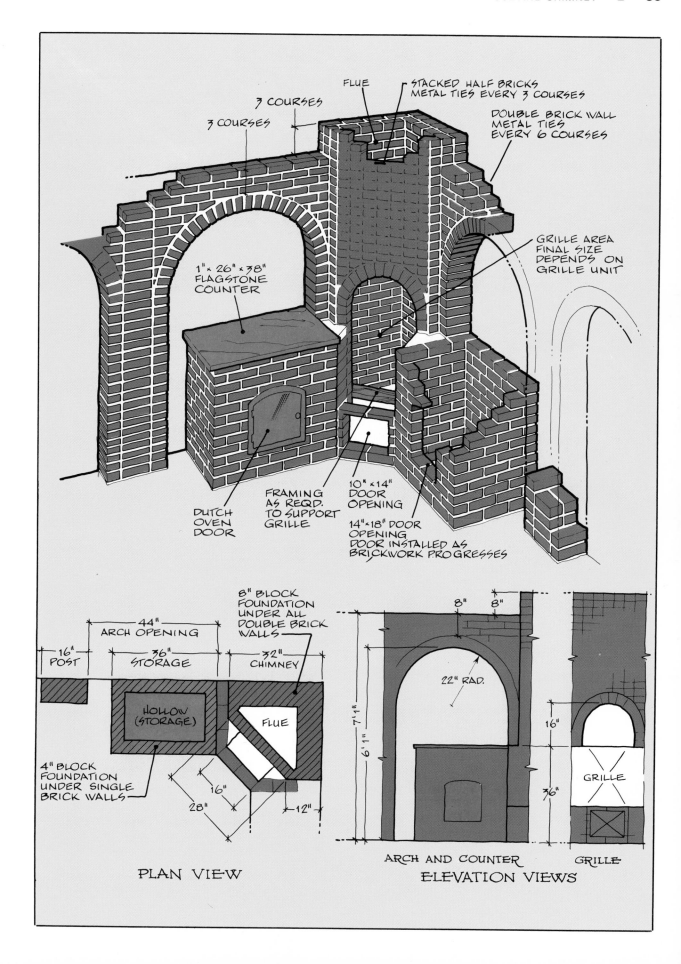

FLUE

3 COURSES

3 COURSES

STACKED HALF BRICKS
METAL TIES EVERY 3 COURSES

DOUBLE BRICK WALL
METAL TIES
EVERY 6 COURSES

GRILLE AREA
FINAL SIZE
DEPENDS ON
GRILLE UNIT

1" × 26" × 38"
FLAGSTONE
COUNTER

FRAMING
AS REQD.
TO SUPPORT
GRILLE

10" × 14"
DOOR
OPENING

14" × 18" DOOR
OPENING
DOOR INSTALLED AS
BRICKWORK PROGRESSES

DUTCH
OVEN
DOOR

8" BLOCK
FOUNDATION
UNDER ALL
DOUBLE BRICK
WALLS

44"
ARCH OPENING

16"
POST

36"
STORAGE

32"
CHIMNEY

HOLLOW
(STORAGE)

FLUE

4" BLOCK
FOUNDATION
UNDER SINGLE
BRICK WALLS

16"

28"

12"

8"

8"

22" RAD.

7'1"

6'1"

16"

36"

GRILLE

PLAN VIEW

ARCH AND COUNTER
ELEVATION VIEWS

GRILLE

The brickwork over the arch and the grille is stacked half bricks with plumb vertical joints. Wall ties should be laid across the joints every third course to bond the half bricks together.

Cut off with the trowel all mortar that protrudes on the inside to keep as smooth a surface as possible. This area can be parged (plastered) with mortar with the trowel to make sure that all of the mortar joints are full.

Continue to build the brickwork, inserting wall ties in the adjoining arch wall to tie all of the brickwork together.

The total height of the wall over the grille is 7'9", as shown on the plan. The adjoining arched opening walls are 7'1". The last course of brick laid on all of the walls should be filled fully with mortar and smoothed off with the pointing trowel to make it waterproof.

After three days have passed, remove all arch forms and point any holes left on the underside of the arch with fresh mortar.

After the brickwork has cured for at least a week, it can be cleaned down with a solution of 1 part muriatic acid to 10 parts water to remove all stains and mortar particles. Rinse with plenty of water from a hose with a nozzle to remove any chemicals and dirt.

This handyman built the walls adjoining the barbecue to meet his existing house walls, thus having it form a patio. A series of arched openings were built to accomplish this. The photograph shows only the brick barbecue, chimney, and a typical arched opening on both sides. You can build it as long or as short as desired.

He also built electrical conduit into the walls on both sides of the arched openings and installed lantern-type electrical fixtures to illuminate the patio and barbecue area at night.

Potted plants placed around the area added the finishing touch. The metal grille units make an efficient cooking surface and can be removed and stored during the winter months.

A trellis of 2×6 lumber overhead with trailing vines completes a relaxing outdoor area with soft lighting from the filtered sunlight. You may not wish to go as far as this handyman did, but the barbecue and chimney will enhance any outdoor area and will be enjoyed by all of the family.

This outstanding project was designed and built by William Sprigg, Frederick, Maryland. He is a professor of music and organist.

HAND TOOLS

Bricklayer's trowel	2' square
Bricklayer's hammer	Ball of line
Mason's modular rule	Convex jointer
4' level	Slicker jointer
2' level	Pair of line blocks
Pointing trowel	Line pin and nail
Bricklayer's broad chisel	Brush

EQUIPMENT

Wheelbarrow	Bucket and hose
Mixing box	Mortarboard
Shovel and mortar hoe	Pick and digging iron

MATERIALS

(*Note:* Concrete for the footings of the barbecue, chimney, and adjoining brick storage cabinets are only estimated here, since the arched openings and walls for your project could vary considerably from the photo. However, estimating factors for brick, block mortar, etc., are given below. For estimating any additional bricks needed for adjoining walls, figure seven bricks per square foot of wall. One bag of masonry cement will lay 125 bricks. One ton of sand will lay 1,000 bricks. Estimate any additional concrete blocks for foundation walls by adding the total lineal feet of wall together and multiply this figure by .75. This is because it takes three blocks to every 4' of length. Figure that 1 bag of masonry cement will lay about 30 blocks. Allow 8" in height for each course of blocks.)

For Footings

Approx. ¾ cubic yard concrete, for the footings under barbecue and brick storage cabinets. (*Note:* I suggest pouring the footing as one complete slab rather than trying to pour for each individual wall. If you mix the concrete yourself, you will need 4 bags of portland cement, 1,100 lbs. sand, and 1,300 lbs. stone.)

For Foundation

14 concrete blocks, 8"×8"×16"
35 concrete blocks, 4"×8"×16"
2 bags masonry cement, regular Type N
500 lbs. sand

For Walls above Grade

Approx. 1,100 standard bricks (allowing about 30 for breakage)
8 bags masonry cement, regular Type N
1 ton building sand
2 flagstones, 26"×37", for tops of brick storage cabinets (*Note:* If you cannot buy them this large, use two pieces for each top with a mortar joint in the center.)
Approx. 40 corrugated metal wall ties, to tie the stacked brick walls together and at the angles

Special Materials

The cast-iron metal doors shown in the brick storage cabinets and the metal outdoor fireplace units are available from different manufacturers and are usually stocked by masonry building suppliers. The particular one shown and used in this project is manufactured by Vestal Manufacturing Company, a division of the Celotex Corporation, a subsidiary of the Jim Walter Corporation, Sweetwater, TN 37874. If you have a problem obtaining one of these units or doors, write the factory for the nearest dealer. The metal fireplace grille shown in the photograph is Model OF-56 of the Vestal line. The cast-iron metal fireplace grille unit is designed to be slid out in winter if so desired. However, if it is given reasonable care the weather will not damage it. The cast-iron doors are known as Dutch oven doors and are also available from Vestal.

28
BRICK FIREPLACE

What makes this fireplace different from the usual one is the cantilevered hearth and brickwork built out on angle irons over the firebox area. The lamps on the ends and the real candles mounted in the center make it a beautiful, creative fireplace that provides a warm, cheery atmosphere. The fireplace is built in the center of the room and it is the focal point of the home.

Start by staking out and pouring a footing for the fireplace and chimney as shown on the plan. Mix the concrete in a proportion of 1 part portland cement to 2 parts sand to 4 parts crushed stone.

Build the blockwork up to the finished floor level as shown on the plan. If you have a basement under the house, you will need more materials.

Build the brickwork up to the height where the concrete reinforced hearth is to be poured. Build a wood form in the front and sides of the fireplace to hold the concrete. Install the reinforced steel rods (rebar) in a grid pattern on the form. See plan for size of hearth.

Before pouring the concrete in the form, lightly oil the form with clean oil (a light motor oil works fine).

Make sure when pouring the concrete in the form that it is vibrated either by a vibrator, which can be rented, or by tapping down on and around the form. This is necessary to achieve a smooth surface on the concrete. The concrete need not be troweled smooth on top of the hearth, as flagstone and brick will be laid on top of it later.

Let the concrete cure about three days before removing the forms. Any holes or voids can then be pointed up and rubbed smooth with some thin portland cement. If the work was done correctly, there probably will be no holes.

Build the brickwork up on the back and sides and leave the opening in the front as shown on the plans.

Lay the firebricks on the fore hearth and inner hearth as shown on the plan. Build the firebox to the correct height and set the poker-type damper in place. Consult the plans as work progresses.

The angle irons are cut at a 45° angle on the corners and welded together to form a three-sided rectangular support. Set this into position and brace so they are level. As shown on the plans, the angle irons should project out about 24″ with the

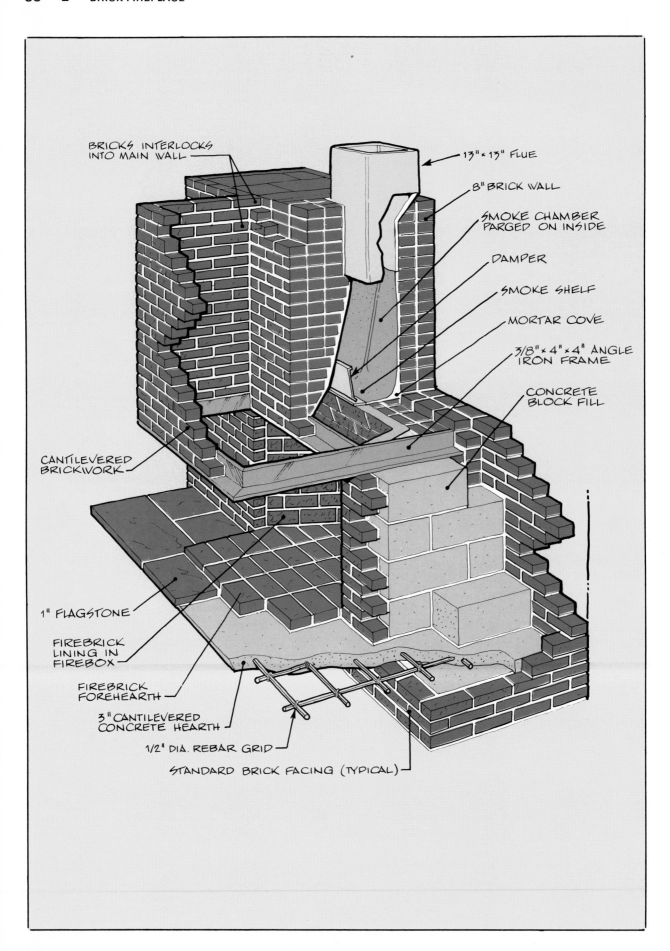

BRICKS INTERLOCKS INTO MAIN WALL

13" × 13" FLUE

8" BRICK WALL

SMOKE CHAMBER PARGED ON INSIDE

DAMPER

SMOKE SHELF

MORTAR COVE

3/8" × 4" × 4" ANGLE IRON FRAME

CONCRETE BLOCK FILL

CANTILEVERED BRICKWORK

1" FLAGSTONE

FIREBRICK LINING IN FIREBOX

FIREBRICK FOREHEARTH

3" CANTILEVERED CONCRETE HEARTH

1/2" DIA. REBAR GRID

STANDARD BRICK FACING (TYPICAL)

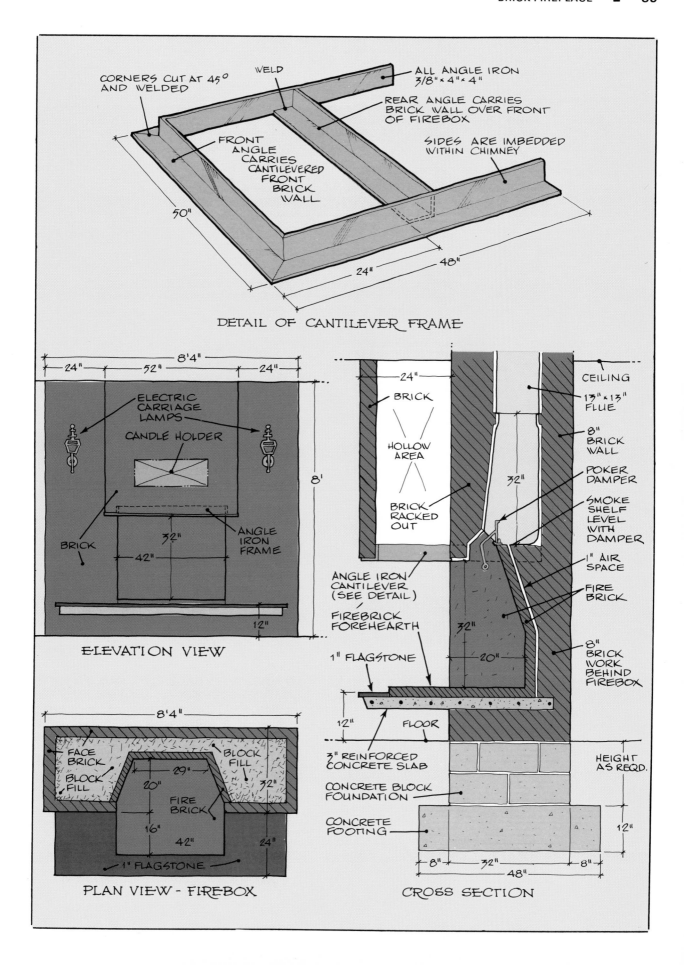

DETAIL OF CANTILEVER FRAME

CORNERS CUT AT 45° AND WELDED

WELD

ALL ANGLE IRON 3/8" × 4" × 4"

REAR ANGLE CARRIES BRICK WALL OVER FRONT OF FIREBOX

FRONT ANGLE CARRIES CANTILEVERED FRONT BRICK WALL

SIDES ARE IMBEDDED WITHIN CHIMNEY

50"

24"   48"

ELEVATION VIEW

8'4"

24"   52"   24"

ELECTRIC CARRIAGE LAMPS

CANDLE HOLDER

BRICK

ANGLE IRON FRAME

8'

32"

42"

12"

PLAN VIEW - FIREBOX

8'4"

FACE BRICK

BLOCK FILL

BLOCK FILL

29"

20"

32"

FIRE BRICK

16"

42"

24"

1" FLAGSTONE

CROSS SECTION

24"

BRICK

HOLLOW AREA

BRICK RACKED OUT

ANGLE IRON CANTILEVER (SEE DETAIL)

FIREBRICK FOREHEARTH

1" FLAGSTONE

12"

FLOOR

3" REINFORCED CONCRETE SLAB

CONCRETE BLOCK FOUNDATION

CONCRETE FOOTING

CEILING

13" × 13" FLUE

8" BRICK WALL

POKER DAMPER

SMOKE SHELF LEVEL WITH DAMPER

1" AIR SPACE

FIRE BRICK

8" BRICK WORK BEHIND FIREBOX

HEIGHT AS REQD.

32"

32"

20"

12"

8"   32"   8"

48"

remainder being built back into the fireplace. Most of the weight and angle are built back into the fireplace so that the overhanging section can be safely supported. You may need to have a local welding shop make this up for you. Be sure to have all the welding beads flattened or hammered off so as not to interfere with brick that will lie on the irons.

Lay the bricks on the angle irons, forming the projected front of the chimney. The area directly back of this will be hollow, as shown on the plans. See section view.

Corbel the smoke chamber in, also as shown on the plan, and set the flue in position 32" above the top of the damper.

Continue building the fireplace up to ceiling height as shown on the plans.

After the brickwork has set about one week, it should be washed down with a solution of 1 part muriatic acid to 10 parts water, making sure that the interior of the house is protected from the acid solution.

The lamp fixtures are installed last and connected to a switch. The candle board mounted with lead anchors to the front of the fireplace should complete the project.

This fireplace project could be built when the house is being constructed or added on to an older house. Careful attention should be given to the plans and good workmanship practiced, because of the complexity of the project and the danger of fire and heat. I would recommend this project only to the more serious handyman who has studied the design and construction of fireplaces and chimneys.

HAND TOOLS

Bricklayer's trowel	Pair of line blocks
Bricklayer's hammer	Convex jointer
Mason's modular rule	Flat slicker tool
4' level	Pointing trowel
2' level	Brush
Brick chisel (broad-edged)	Chalkbox
Ball of line	

EQUIPMENT

Mixing box	Mortarboard
Shovel and hoe	Wheelbarrow
Bucket and hose	

MATERIALS

(*Note:* This fireplace was built on a concrete footing in a house that did not have a full basement. All that is necessary is that the footing and foundation be built below the existing freeze line for your local area. If there is a basement area, allow enough additional materials for this work. The same is true for the chimney above the ceiling level of the room. The plan concerns itself only with the construction of the fireplace and the part of the chimney which is exposed inside the room of the house, as the chimney is of a typical construction.)

For Footing

1 cubic yard concrete (*Note:* This is allowing for a depth of 12" with an 8" spread more than the actual walls for distribution of weight. In addition, the footing should be reinforced with steel rods, concrete reinforcing wire, or metal pipe.)

56 concrete blocks, 8"×8"×16", for two courses of concrete block from the footing to the finished floor

2 bags masonry cement, regular type

Approx. ¼ ton sand

For Walls above Finished Floor

Approx. 2,000 standard bricks, for facing of fireplace and interior work around smoke chamber and flue area

40 split firebrick, 1¼"×4½"×9", for outer and inner hearth (*Note:* These are readily available at masonry materials suppliers.)

60 regular firebricks, 2½"×4½"×9", for firebox walls

1 poker control damper, 42" long

3 angle irons, ⅜"×3½"×50", welded together at the corners to form three sides of a rectangle and serve as a support to carry the brickwork over the firebox area (*Note:* See detail on plan of how they are welded together.)

Approx. 75 concrete blocks, 8"×8"×16", for fill on both sides of the firebox and flue area of the chimney (*Note:* This is based on a fireplace 8' in height to the ceiling of the room. You can substitute old used bricks or masonry rubble if available.)

Flue linings, 13"×13" (*Note:* Number needed will depend on how high you build your chimney. The first one should be set in position about 32" above the top of the damper as shown in the section view.)

Approx. 2 bags portland cement, 400 lbs. sand, and 500 lbs. crushed stone (about ¼ cubic yard), to pour the cantilevered concrete hearth for the fireplace.

Approx. 8 reinforced steel rods, ½"×48", to reinforce the cantilevered hearth

Approx. 7 steel rods, ½"×7' to form a grid (*Note:* Grid should be wired together lengthwise.)

Plywood and lumber to support concrete cantilevered hearth in the front and sides where it projects until it has hardened

Approx. 16 square feet flagstone, for top of outer hearth

Approx. 19 bags regular masonry cement, for all masonry above finished grade

Approx. 2½ tons building sand

2 electric lamps (*Note:* These are optional. If you use them, be sure to install conduit pipe for the wiring as the masonry is built. Also keep the pipes away from the hot area of the chimney.)

Candle board or mount (*Note:* This is optional. The fireplace in the photo features real candles mounted on a piece of hammered brass. The builder of this fireplace considered this one of the most important features.)

29

STONE DRIVEWAY ENTRANCE WALL

tart by staking out and excavating the footing areas. Make sure it is below the existing frost line for your area. Mix and pour the footings in a proportion of 1 part portland cement to 2 parts sand to 4 parts crushed stone.

The footing for this project was poured on a sloping area, so it was stepped to save materials, as shown on the plan view. If you do this as shown, overlap the concrete steps about 12″ for a good bond. If there is no slope, then pour the footings level.

Lay the blockwork up to grade line, stepping it up over the footings. Since the wall is 24″ wide, two 12″ concrete blocks were used, laying them side by side. The piers on the ends are 3′ wide and required some 6″ blocks along with 12″ blocks to make the foundation wide enough.

The stones used on this project came from a stone quarry and were fairly straight on the tops and bottoms, requiring mostly that they be cut to length and dressed on the corners. As the mortar sets up between the stones, it will safely support the weight of the stone. In this project it was raked out

about ½″ with an old broom handle and brushed after it was dry enough not to smear.

The piers were built on each end first and then the wall laid to a line in between. Notice that there is a setback of the wall from the piers. This is because the piers are 3′ wide and the wall only 24″ wide.

The center of the wall and piers was filled in with rubblestone and mortar, making it solid.

When the top of the piers and wall was reached, flat thin stones were picked out and laid for the cap. Let them project about 1″ on all sides to serve as a water drip and highlight the stonework. All mortar joints on the top of the wall should be tooled flat with the slicker tool or pointing trowel to keep out water.

If you install electrical lamps in the wall, be sure to build the electrical conduit pipe into the wall as the work progresses. This pipe is available from any electrical supply house.

Brush off all the work to remove any mortar particles and dirt when completed. Repoint where necessary. After the stonework has set for about a week, it can be washed down with a solution of 1

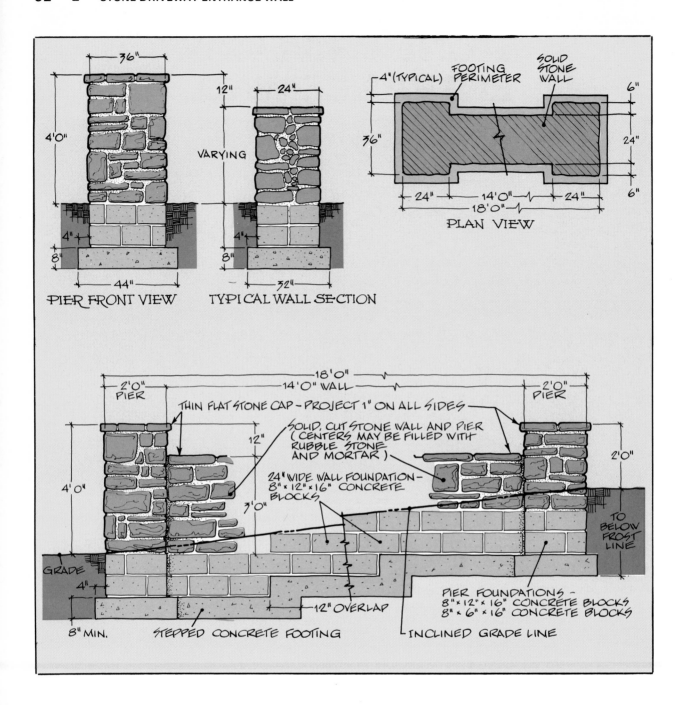

PIER FRONT VIEW   TYPICAL WALL SECTION

PLAN VIEW

part muriatic acid to 10 parts water, rinsing with plenty of clean water from a hose with a nozzle attached.

Fill in around the wall with earth and topsoil and plant some flowers or shrubbery to complete the job.

A stone driveway entrance wall will not only give your property a stately appearance but provide a safe entranceway at night. The stonework will last for many many years with little or no maintenance.

HAND TOOLS

Bricklayer's trowel
Bricklayer's hammer
Mason's modular rule
4' level
2' level
Pointing trowel
2 stone chisels (one broad-bladed, one pointed)

2' square
Ball of line
Line pin and nail
Flat slicker tool
Rounded-off piece of broom handle, to rake out mortar joints
Brush
Steel tape, 50'

EQUIPMENT

Wheelbarrow
Mixing box
Mortarboard

Shovel and hoe
Bucket and hose
Pick and digging iron

MATERIALS
For Footings

(*Note:* There are two stone entrance walls, one on each side of the driveway. Materials listed are for only one wall. Double all materials if you desire a wall on each side of driveway.)

1½ cubic yards concrete (*Note:* This would be 8" deep and include the wall and both of the piers shown.)

For Foundation Walls

54 concrete blocks, 12"×8"×16"
12 concrete blocks, 6"×8"×16" (*Note:* These are needed for the piers in addition to the 12" blocks as the piers are larger than the wall.)
2 bags portland cement, 2 bags hydrated lime, and approx. 500 lbs. sand to mix Type N mortar. (Formula for Type N: 1 part portland cement to 1 part hydrated lime to 6 parts sand.)

For Walls above Grade

● Estimating the amount of stone will vary as explained in previous projects depending on the type of stone being used. Find out what 1 cubic foot of the stone you are going to use weighs by weighing it on a scale. Then calculate the square footage of stone in the wall and convert it into cubic feet. This will give you the amount of stone needed. A ballpark figure of 125 pounds per cubic foot will be close. If you have free access to the stone from an old fence, barn, stone wall, etc., then there would be no need to estimate it, as all it will cost you is the hauling expense.

● As was the case with the amount of stone needed, the amount of mortar will depend on how many voids there are between the stones and how much filling in is required. Use the Type N mix described for the foundation walls. I suggest buying a couple of bags of portland cement and lime and seeing how far it goes. This will give you a basis to work from. Start also with about 1 ton sand and do the same as for the cement and lime.

● The electrical lamp and fixtures are optional. Electrical conduit pipe should be built into the walls as they are constructed to protect the wiring. The same is true for the electrical box that the fixture will be mounted to. If you do not have any wiring experience, have an electrician install that part of the job. A switch to turn the lights off and on should be located inside the house or garage.

30

WATERFALL AND FOUNTAIN

This is one of the more creative and enjoyable projects in the book. I hope you will enjoy building it as much as I did. When I designed my patio, which adjoins it, my wife and I felt that a corner of it would be the ideal spot for a pool. It would make a perfect place to unwind after a hard day's work and would be a source of enjoyment for the grandkids.

After sketching several ideas on paper, we decided to build a brick border wall in back of the waterfall to serve as a backdrop and support for the stonework. The waterfall and fountain would be of rubblestone and laid as naturally as possible.

Start the construction process by staking out and excavating for the walls and pool area as shown on the plan. We wanted it deep enough to support a few goldfish and water plants. The fish would control the insect population.

Next, I poured the footings as shown on the plan around the perimeter of the foundation.

After waiting a day to allow the concrete to harden, two courses of blocks were laid in mortar. Type S mortar is relatively waterproof and is excellent for a project such as this. If you cannot obtain Type S masonry cement, substitute by mixing 1 part portland cement to ½ part hydrated lime to 3 parts sand, with water. After the blocks were laid, I parged the exterior side with mortar to make it more waterproof. A few strips of wire mesh were laid in the mortar bed joints and left projecting out so that they would tie into the concrete side walls when poured.

The brickwork was built up to the top of the walls and a rowlock cap laid on top to finish this part of the project.

The next step was to build the plywood form for the concrete lining walls. I made a square form of ⅝" plywood, 24" high, and nailed 45° angle braces at the corners. The concrete was mixed for the bottom first, to a proportion of 1 part portland cement to 2 parts sand to 3 parts gravel, with water.

Lay reinforcement wire on the pool floor for strength. Pour and trowel the pool bottom smooth, making sure that the wire is well embedded in the concrete.

Then I immediately set the wood form in position on the wet concrete and poured the sides, lining the walls to form a watertight unit. After the concrete was poured in place, I tapped around the sides of the form with a hammer lightly to prevent any

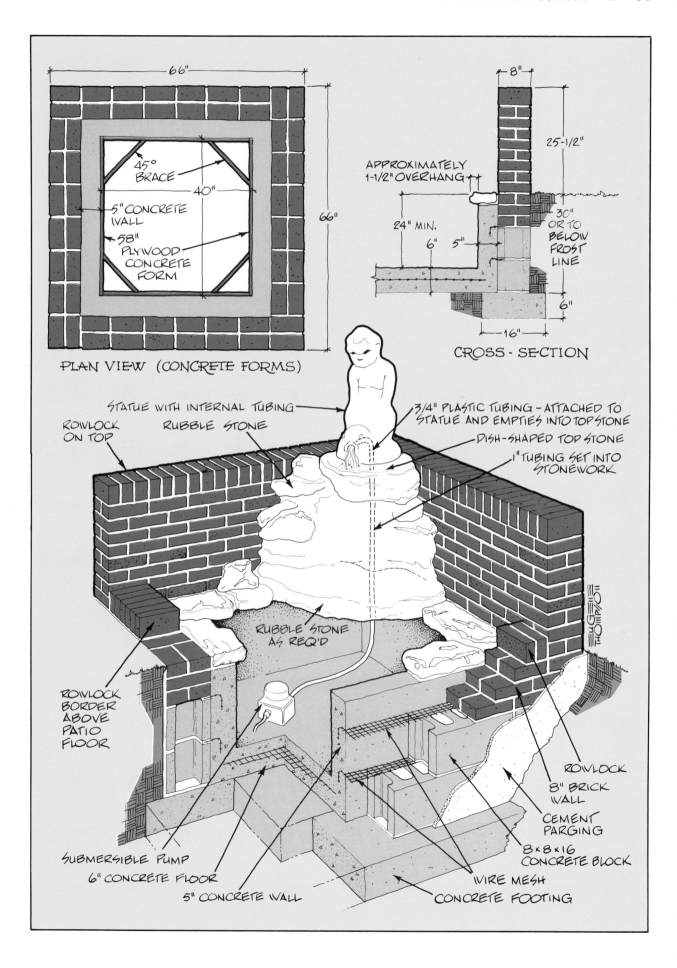

PLAN VIEW (CONCRETE FORMS)

66"

45°
BRACE

40"

5" CONCRETE
WALL

58"
PLYWOOD
CONCRETE
FORM

66"

CROSS - SECTION

8"

25-1/2"

APPROXIMATELY
1-1/2" OVERHANG

24" MIN.

6"   5"

30"
OR TO
BELOW
FROST
LINE

6"

16"

STATUE WITH INTERNAL TUBING

ROWLOCK
ON TOP

RUBBLE STONE

3/4" PLASTIC TUBING – ATTACHED TO
STATUE AND EMPTIES INTO TOP STONE

DISH-SHAPED TOP STONE

1" TUBING SET INTO
STONEWORK

RUBBLE STONE
AS REQ'D

ROWLOCK
BORDER
ABOVE
PATIO
FLOOR

ROWLOCK

8" BRICK
WALL

CEMENT
PARGING

8 x 8 x 16
CONCRETE BLOCK

SUBMERSIBLE PUMP

6" CONCRETE FLOOR

5" CONCRETE WALL

WIRE MESH

CONCRETE FOOTING

voids or air bubbles from marring the surface when the forms were removed later. It is also a good idea to lightly oil the inside of the forms to make a smoother finish.

I allowed the concrete to cure about two days before removing the forms.

The stonework was started by laying the cornerstone that forms the base of the waterfall first and then working in the others around the concrete edge of the pool. Try to select the base stone for the waterfall that is triangular and large enough to span the corner. See plan for information on this. It should be a rather flat stone.

The stone ledge around the inside of the pool should project about 1½" to hide the concrete edge and permit the water to go back slightly under the stones when filled.

A hole was left at the rear center of the angled waterfall stone for the plastic line coming from the pump, which would be installed when the pool is complete. I inserted a 1" plastic tube, long enough to carry the water from the bottom of the pool to the top. Plastic tubing is available from almost any hardware store. As the stonework was built, I walled the plastic tubing in securely. See plan for a view of the plastic tubing in stonework.

The stonework in the waterfall was laid in no particular pattern but with each succeeding layer setting back over the one beneath to form a pyramid effect and resemble a natural waterfall. A few stones set out farther than the others, forming a ledge to place objects on, such as a frog, etc.

I selected a dish-shaped top stone to provide a plate for the water to collect before running down over the stonework. To make mortar joints look natural, I used a short length of old broom handle to rake out the mortar slightly and finished by brushing after the mortar had set enough not to smear.

I bought a small circulating pump from a Sears garden shop and a supply of ¾" plastic tubing to hook up to the end of the pump. This ¾" tubing I inserted through the original 1" tubing built into the stonework. This acted as a sleeve for the ¾" tubing, and if the ¾" tubing ever deteriorates I can replace it with a new section. I obtained the concrete figure that forms the top of the fountain from a local garden shop that stocks these items. The plastic tubing was inserted through the hole in the casting of the figure, completing the installation. Make sure that it fits snugly so water does not leak out at the bottom and prevent a full spray of water.

The pump should have a plastic protective screen where the water intake is to prevent clogging with leaves or dirt. The Sears pump has this feature. The pump operates off regular household current. Occasionally water has to be added to replace water lost through evaporation.

In cold weather, the pump should be disconnected and drained for storage. In addition, I toss a short length of firewood or log in the pool to absorb the pressure created by the freezing and thawing of the water, which otherwise could crack the walls of the pool.

HAND TOOLS

Bricklayer's trowel	Nail and line pin
Bricklayer's hammer	Flat slicker jointing tool
Pointing trowel	Short piece of old broom
Mason's modular rule	handle for raking out
2' level	and smoothing mor-
4' level	tar joints in
2 chisels (wide-bladed	stonework
and narrow-bladed)	Brush
Ball of line	Tin snips

(*Note:* In addition, you will need some basic carpentry hand tools such as a saw, square, claw hammer, etc., to build the plywood form required for the inside of the pool.)

EQUIPMENT

Wheelbarrow	Bucket and hose
Shovel and hoe	Mortarboard
Mixing box	Pick and digging iron

MATERIALS
For Footings and Pool
Approx. 1 cubic yard concrete
Piece of concrete reinforcement wire, 52"×52" for bottom of pool

For Foundation Walls
38 concrete blocks, 8"×8"×16"
2 bags Type S masonry cement
300 lbs. sand
Metal wire mesh, commonly called hardware cloth, to be projected from the bed joints of the blockwork to reinforce the concrete sides of the pool when it is poured

For Walls above Grade
Approx. 500 standard bricks
4 bags Type S masonry cement
1 bag portland cement
1 bag mason's hydrated lime
¾ ton sand
Approx. 2 wheelbarrows rubblestone, for waterfall and fountain, including edges around the pool (*Note:* Try to select a nice triangular stone that is rather flat for the corner, as it has to span that area and rest on the stone edging work. Visit a site in your area where there is an old stone barn, bridge remains, or rock outcropping for stone. Old fence rows are also a good source of rubblestone.)
Plywood, ⅝"×¾" thick, to serve as forms for the sides of the concrete pool lining

Part II
MASONRY TECHNIQUES

Masonry work—brick, block, concrete, and stone—reflects a long and honored tradition of progressive attitudes and pride of accomplishment. It is one of the oldest trades, but most of the techniques have changed little over the years. The aim of Part II is to describe, in detail, the necessary techniques to complete the masonry work contained in Part I. The techniques covered in the following pages can also be used to do other masonry projects around the home. The last chapter of Part II details how to maintain or repair all types of masonry work.

31
BRICKLAYING: TOOLS AND TECHNIQUES

his section should serve as a reference for building the brick projects in this book. Here you'll find a complete list, with brief descriptions, of the bricklayer's tools, an illustrated course in basic bricklaying, and supplementary data on estimating and buying materials. Also included are numerous tips and time-saving methods that I've learned over the years. Some of them I devised myself; others I learned from other masons. All of them will give your work a professional finish.

Keep in mind, however, that there is usually more than one way to do anything. The projects in this book were built, for the most part, by amateur masons. Each one has his own style and technique. You'll have to find your own way, too, after you've mastered the basic tool skills.

Take your time and enjoy the creative pleasure of bricklaying, of constructing something solid and enduring and attractive. Don't measure your work by how fast you do it, but on the finished job. If there's one talent a bricklayer ought to possess it's patience.

TOOLS YOU WILL NEED

The hand tools you will need to lay brick and block are rather few in comparison to other types of work. Since this is true, I would recommend you buy only tools with well-known brand names, as they will last longer and handle better. You can obtain these tools at most reputable hardware stores or building supply houses. You can rent the more expensive hand tools or equipment that will not be used frequently—check the yellow pages of your telephone book for the nearest rental store to you.

If for some reason you have a problem locating masonry tools, there are two excellent mail-order companies I recommend. Both of these firms will send you free catalogs if you write to them. Most of the professional masons in my area buy from them. They are as follows:

Masonry Specialty Co.
4430 Gibsonia Road, Rt. 910
Gibsonia, PA 15044
Telephone (412) 443-7080

Goldblatt Trowel Trade Tools
511 Osage, P. O. Box 2334
Kansas City, KS 66110
Telephone (913) 621-3010

For each of the projects in this book, I have listed, as much as possible, the tools and materials needed. Following is a summary list of those tools with a few tips about selecting them.

Recommended Masonry Tools

Bricklayer's trowel
London-pattern blade, good-size (10½" to 11"), is best.

Pointing trowel
Average blade size is 5"×2½".

Brickhammer
18 oz. is a good all-around weight.

Levels
48" length is best for most work. I would buy one with alcohol vials. If you want to invest in a smaller level for cramped places, a 24" size can be helpful.

Brick set chisel
This broad-edge chisel is the best tool for cutting bricks. Average size is 3½"×7".

Standard mason's cutting chisel
This is good for cutting out old work, cinder block, and some veined stones. A good size is 2¼"×7½".

Convex jointer
Most often used jointer for half round rodded or sunken joints. Best average size is ⅝"×¾". Also available in sled-runner type.

V jointer
Makes a V impression in mortar joints; ¾" to ⅞" width. I recommend you buy only the V sled runner for all work.

Flat slicker jointer
Used for pointing flat joints. Best average size is ½" wide at one end and ⅝" wide at the other. Available in different widths or one size with wood handle.

Skatewheel rakeout jointer
This cartwheel rakeout jointer has adjustable

The better the quality of the hand tools you buy, the longer they will last and the easier they will be to use.

thumbscrew for different depths. A must for rakeout joints.

Modular spacing rule

Six scales in 16″ modules on one side, foot and inch graduations on the other side. Overall length is 72″. Standard rule for laying most brickwork and block.

Course counter spacing rule

Ten different scales on one side, standard rule on the other side, 72″ long. Needed when standard course spacing will not work, as in windowsills, arches, etc. You will need one for some of the projects.

Steel measuring tape

A good average length is 50′.

Ball or tube of nylon line

I recommend you buy #18 braided line on a 250′ tube for most work.

2′ steel square

Does not have to be an expensive square, only sturdy, as you will not be concerned with the carpentry tables on it.

Brush

Medium-soft bristle works fine. An old floor brush sawed in half is great.

Chalk box

Most have 100′ of line.

2 line blocks (wood or plastic)

Used to attach line to wall. Most masonry supply houses will give them to you free for asking. They advertise their names on the blocks.

Steel line pins

Also used to attach a line to a masonry wall. Free for the asking at most masonry supply houses.

Pair of work gloves

Leather-palm gloves hold up best.

Tin snips

You need them for cutting metal bands on brick cubes and miscellaneous tasks.

Yellow marking crayon

Handy for laying out. They are called lumber crayons.

Safety goggles or glasses

I recommend you wear safety eye protection when cutting brick or blocks. Glasses are more comfortable than goggles. Side guards on glasses are also recommended.

Recommended Masonry Equipment

Metal-body wheelbarrow

I like the contractor's 5-cubic-foot size, as it holds more mortar and is easier to mix in. Any size will do, however.

Shovel

I recommend one round-point dirt shovel and one square-blade shovel for cleaning up.

Mortarboard

You can make one from a 24″×24″ scrap of ⅝″ or ¾″ plywood. Two 2×4 runners nailed on the bottom provide a base.

Mortar hoe

There are large and small mortar hoes on the market. I like the smaller size as it is easier on the back.

Hose with nozzle

A 50′ hose should be plenty. Great for cleaning up equipment.

5-gallon buckets

You need several of these. White plastic ones used for spackling and industrial cleaners are fine and are usually free for the asking.

Mixing box

Metal ones are available from building supply stores, or you can build your own from ¾″ plywood. A good size is 48″×24″ with 12″ sides and ends. Use coated nails to put it together.

Pick and digging iron

You will need these for most of the projects to excavate for footings and foundations.

Drum-type cement mixer

Although this is not a must, it sure is a back saver when mixing mortar or concrete for small jobs.

Many of the tools and pieces of equipment you probably have now around the house. If not, all of them are stocked at building supply houses. In fairness to manufacturers of masonry tools and equipment, I have not mentioned specific brands. Consult your local supplier or hardware store for recommendations.

BUYING MASONRY MATERIALS

Your best source of materials is the building supply houses that cater to professional builders and masons who want to save money, which is the name of the game. They will generally have a wider selection and will give you a better quantity price. In many cases, they will also take back excess materials for a small handling charge.

Choosing Bricks

There are a number of different sizes and types of bricks. The five most popular sizes are shown in the accompanying illustration.

Of all the sizes, the standard brick is by far the most commonly used. The reasons are simple—standard bricks are more readily available, come in

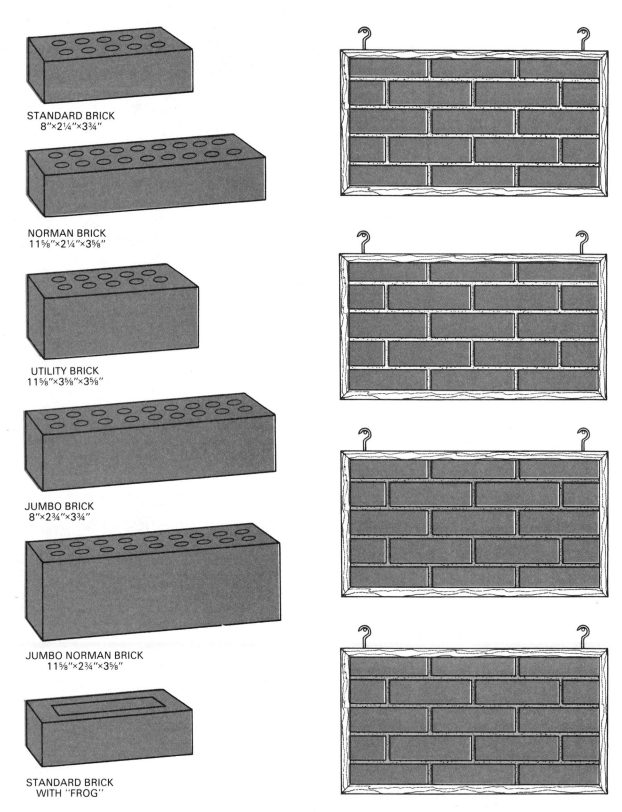

STANDARD BRICK
8″×2¼″×3¾″

NORMAN BRICK
11⅝″×2¼″×3⅝″

UTILITY BRICK
11⅝″×3⅝″×3⅝″

JUMBO BRICK
8″×2¾″×3¾″

JUMBO NORMAN BRICK
11⅝″×2¾″×3⅝″

STANDARD BRICK
WITH "FROG"

There are many different sizes of brick besides those shown here. However, the standard brick, measuring 8″×2¼″×3¾″, is by far the most commonly used.

Sample boards of bricks demonstrate how each type will look when laid in a wall, though you may plan on using a more interesting bond.

a greater range of colors, and are less expensive. They are also made in many different textures and finishes.

The best method to use for selecting or choosing a brick is to visit a supplier's showroom and look at the sample boards. Various colors and textures will be displayed with a simulated mortar joint on a small panel or wall section, with the mortar joints tooled (struck) in different popular finishes. You can see exactly how they will look when laid in the wall.

As a rule, the bricks are laid with half laps and have a manufacturer's identification number, which is used for ordering purposes. These are called range or blend numbers. Be sure to jot down the number you want so there is no mistake in your order. It is also a good idea to ask the salesman how long it will be before they are available, as some bricks are made only in kiln runs at certain times of the year.

Bricks are sold singly, in cubes of 500 each, or by the thousand. Naturally, the more you buy the better the price should be. Delivery charges can be high, depending on how far your job is from the supplier. Check into this before ordering. You may want to consider hauling your own if a truck is available and the load is not too big to be impracti-cal. Many firms have a minimum hauling charge, and for a small order the fee could be as much as what the materials cost.

The average brick weighs about 4 pounds, which means that a cube of 500 bricks weighs about 1 ton. Don't overload your truck! It is better to make a second trip than to break down along the road somewhere or be fined for overloading. ·.

It is also a good idea to ask the supplier whether additional bricks will be available if you run out. Some bricks are only made on a limited basis and then discontinued. You may want at some later time to make an addition to your project, and it could be very frustrating if the brick you have selected is discontinued.

You should also decide if you want a brick that is solid, is cored, or has a "frog." Cored bricks have holes that allow the mortar to lock them in secure-ly. Bricks with frogs—an indented area in the bottom—also are locked in well. Both types form a good strong wall. Solid bricks are preferred for paving work, where you don't want any holes, and are also okay for building walls. The choice is up to you. However, if you buy brick with holes, make sure that solid-end bricks are available for the ends of your walls, windowsills, wall caps, etc. Normally they are, but you should ask.

Solid brick, cored brick, and brick with a so-called frog. For many projects any of the three types will do, but cored and frogged bricks do permit especially good mortar joints.

Mortar for Brickwork and Blockwork

The same mortar used to lay bricks is okay for blockwork. There are, however, a number of different types of mortars that can be used.

Unless you're building a small project, I do not recommend buying ready-mixed, packaged mortars that contain all of the ingredients except water, because they are expensive. Remember that the majority of any mortar mix is the aggregate or sand.

For small pointing jobs or where you need only a wheelbarrow or so of mortar, then the ready-mixed mortar may be practical. Many of the stores that stock do-it-yourself materials carry only packaged dry-mix mortars, so you may have to visit a regular building supply house to obtain other types of mortars.

The most popular type of mortar mix sold for masonry work is masonry cement. It is fairly simple to mix and available in different types for specific strength requirements. The general-purpose masonry cement is known as Type N, and unless you specify, this will always be the type sold to you. It is packaged approximately 70 pounds to the bag, and all that is necessary is to add sand and water.

The following proportions are recommended for brick and blocks: 1 part masonry cement to 3 parts sand, with water. To mix what is known as a half batch, mix 9 shovels sand to half a bag masonry cement, with water. To mix a full batch, mix 18 shovels sand to 1 full bag masonry cement, with water. A "shovel" always means a regular round-point dirt shovel. A square shovel would hold too much sand, which results in a weak mortar. If a slightly richer mix is desired, try 16 shovels sand to a full bag masonry cement. The coarseness of the sand may sometimes require a slightly richer mortar mix. You can tell this by how well it sticks to the trowel blade.

One of the most frequent and important jobs when doing your own masonry work is the mixing of mortar. Although this seems to be a simple task, remember that your wall will be no stronger than the mortar that bonds it together.

The following procedure is for mixing masonry cement mortar, which is the most frequently used type:

1. Start by adding 9 dirt shovels of sand to the wheelbarrow, spreading it evenly on the bottom of the bed.

Ready-mixed mortars are very convenient to use but are relatively expensive. However, they are widely available, and the expense is acceptable if the job is small.

General-purpose masonry cement is mixed with sand and water to make mortar for brickwork and blockwork.

Step 1

Step 2

Step 3

Step 4

Step 5

Step 6

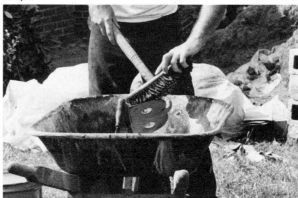

Step 7

2. Next, add one-half bag of masonry cement or 3½ shovels, which are approximately the same amount, if you don't trust yourself to guess at the half-bag amount.

3. Dry mix the sand and masonry cement with the hoe or shovel until blended together. Pull the dry mix to the front end of the wheelbarrow, leaving the opposite end free to hold water.

4. Add about half the amount of water you figure will be needed for the mix. This amount of water will depend on the moisture in the sand or the stiffness you desire. After some practice, this is easily determined.

5. Chop the dry mix with the hoe into the water, mixing the mortar until it is of the desired consistency or stiffness, adding the balance of water as needed.

6. A simple test can be done to determine if the mortar is sticky or rich enough. Pick up some of the mortar on the blade of the trowel and with a downward jarring motion of the wrist, set the mortar on the trowel. Then, turn the trowel upside down. The mortar should stick to the trowel without falling off. If not, add an extra shovel of masonry cement.

7. Last but not least, make sure you wash the mixing tools and the wheelbarrow with water and a brush immediately after mixing to remove all of the mortar. Any mortar left on tools or equipment will dry and pit the steel, causing future mixes to stick and make your mixing job a lot more difficult.

A good rule to follow when making any mortar is not to mix any more than can be used in one hour.

Mortar can be remixed with water to make it workable if it stiffens, but this should be done only once, as tempering (remixing with water) a couple of times will seriously weaken the strength of the mix.

You can stretch your mortar dollar by designing and mixing your own, using portland cement, hydrated lime, and sand. All of these materials are stocked by the building suppliers. To mix portland cement for mortar, use the following proportions for an average mix: 1 part portland cement to 1 part hydrated lime to 6 parts sand, with water. This is equal to a standard masonry cement mortar mix. The addition of lime to the mix gives it a good bonding quality found in few other types of mortars. I prefer always using this mix instead of masonry cement for stonework, because it sticks to the stone much better and is more economical.

To mix a half batch of portland cement and lime mortar, use 4 shovels portland cement to 4 shovels mason's hydrated lime to 21 shovels sand, with water. To mix a full batch, mix 1 full bag portland cement to 1 full bag mason's hydrated lime to 42 shovels sand, with water.

Sand is a very important part of any mortar mix but is often neglected. Before buying sand, ask your building supplier if the sand is washed to remove silt and loam. When silt or loam is present in sand, it prevents the cement in the mortar mix from completely surrounding the sand particles, resulting in a weaker or inferior mix. It also causes the mortar to be sticky and gummy—hard to handle on the trowel—and could cause an off color. Mud or silt particles could cause tiny holes to appear in the mortar joints after the mortar hardens. As a rule, most sand sold by suppliers has been washed, but

You can make your own masonry mortar by mixing equal parts of mason's hydrated lime and portland cement with sand and water. This mix bonds well and is especially good for stonework.

If a handful of sand remains in a lump after you have squeezed it, the sand probably contains too much silt or loam to make a good mortar.

you should inquire. Be especially wary of buying sand that comes directly from a sand pit if there is one in your area, as it will usually be too dirty to use.

There are two simple tests you can perform to determine if the sand is clean. One is to pick up a handful and squeeze it into a ball. If the sand remains in a ball when you open your hand, it probably contains too much silt. The sand should be fairly dry, as wet sand would stay in a ball even if there was no silt present.

The second is called a "siltation" test and is very easy to perform, If silt, clay, or impurities are suspected in the sand do the following: Fill a glass jar (either pint or quart size) half full of sand. Add about 3" of water. Shake the jar vigorously and let it set overnight. The sand will settle to the bottom of the jar and the silt or impurity will rise to the top of the sand. If the accumulation of silt or impurities is more than ⅛", the sand should not be used for masonry mortar. The illustration below shows the silt on top of the sand. There is in excess of ⅛" of silt and this sand should not be used.

You can get the best price on sand by hauling it yourself in a pickup truck. Many builder supply houses have a minimum charge for delivery. If you are going to order more than 2 tons or can have other materials delivered at the same time, it may be practical and economical to have it delivered by the supplier. But the cost of hauling a small load of sand could be as much as the sand itself is worth.

My last point on sand is very important! Mortar sand is much finer than concrete sand. Concrete sand has small pieces of gravel in it and is much coarser. Mortar made with concrete sand for laying brick or block is extremely difficult to work with and sets very fast in the wall. Be sure to tell your building supplier that you do not want concrete sand for your mortar.

HOW TO ESTIMATE BRICKS AND MORTAR BY RULE OF THUMB

I would like to emphasize that rule-of-thumb estimating is not intended to be a perfect mathematical method of figuring a job. It always allows a little

Silt rises to the top of the sand; if the accumulation is more than ⅛", as shown here, it should not be used.

for waste. If one is estimating for a large building or apartment complex, then established tables and a slide rule or calculator would be used. Rules of thumb are, however, very workable and reasonably accurate for jobs up to as large as a house.

Rule-of-thumb estimating was developed over the years from practical experience of figuring the job and then being around to see whether the figuring works out. It is not necessary to add an allowance for any waste, as this is already figured in. Even though a material list is supplied for each project wherever possible throughout the book, there may be times you want to enlarge or reduce a project. Refer to the information in this section as a guide. Rules of thumb for estimating blockwork and stonework will be given in the appropriate chapters.

Estimating Brick

First, find the amount of square footage of the wall or project. This is done by multiplying the height by the length. Round off fractions to the next higher number, as it won't make that much difference. The information on brick is referenced to the standard brick, of which most projects in this book are built. Brick of smaller or larger size than standard will require an adjustment in figures.

For example, if you are building a brick wall that is 5' high by 10' long, it would take 50 square feet of brickwork. This is for a single thickness of wall or what is called a 4" wall. If you have an 8" wall, then the figures have to be doubled. If there are any openings, the square footage of them would also have to be deducted. Mortar joints are included in the square footage; do not allow any additional for them. Generally, the standard thickness of mortar joints will be ⅜" for brick or block. Old used brick will usually require a ½" mortar joint because they are less uniform. The difference, however, is minor and no extra allowance needs to be made.

To determine how many bricks will be needed for the 50 square feet of wall area, multiply by 7. This will allow for a small amount of waste, such as cracked or broken bricks. Technically speaking, there are exactly 6.75 bricks in a square foot of wall. Therefore, there would be 350 bricks in the example wall above. This figure is of course for a single width brick wall.

Rule-of-Thumb Formula for Standard Brick:
Square feet of wall area (minus square feet of any wall openings) × 7 = quantity needed

Estimating Mortar

One bag of masonry cement combined with 18 shovels of sand will lay about 125 bricks. Based on this, using the example above, divide the 350 bricks by 125 to determine how much masonry cement you need. This comes out to 2.8 bags. Since you can't buy a part of a bag, it will take 3 bags.

If you are using a dry packaged mix that only needs water added to make mortar, figure that each 40-pound bag will lay about 50 bricks. Most packaged mortar mixes weigh about 40 pounds.

If you decided to use a portland cement and lime mortar mix, figure that a full batch, which is 1 bag portland cement to 1 bag hydrated lime to 42 shovels sand, will lay approximately 300 bricks. This is a lot of mortar to prepare at one time, so unless you have a lot of help, I recommend mixing only a half batch at a time. As far as cost is concerned, portland cement and lime mortar mix is more economical than other types, but it does take more time to prepare.

Estimating Sand

Sand is by far the least expensive mortar material you have to buy. To estimate the amount of sand needed, allow 1 ton of sand for every 8 bags of masonry cement needed. This amount will lay about 1,000 bricks.

Remember that when you buy sand, some of it will be lost on the ground. Therefore, I would buy an extra ½ ton if you need in excess of 2 tons. It never goes to waste; you can always put it in the kids' sandbox or somewhere around the grounds. If the weather is cold or freezing, the sand should always be covered to keep moisture out.

TIPS ON STORING MATERIALS

When your materials are delivered, make sure that they are set on wood boards off of the ground and covered until you're ready to use them. This is true in warm weather as well as cold! Dampness, water, or frost causes many problems if it gets into the bricks. If the bricks are too wet, they will not absorb the moisture from the mortar as they are laid, resulting in sloppy smeared work and a lot of difficulty in keeping them plumb and level. In addition, some bricks have a natural salt in the clay, and when this salt unites with an excessive amount of moisture, it will cause a white scum or deposit to appear on the face of the bricks. This is known in the trade as efflorescence. The only way to remove this stain is by wetting the wall with a hose and scrubbing well with a stiff brush with a solution of 1 part muriatic acid to 10 parts water. Then flush the wall off again with water from a hose with a nozzle attached, to remove all of the dirt and chemicals. The stain may reappear until the bricks have completely dried. This is why you see brick and masonry walls covered on construction jobs, even in warm weather.

The same is true for cements. If they get damp or wet, the process of hydration occurs—that is, the cement hardens. Mixing this partially hardened cement results in a mortar that will have hard lumps in it, which is not only frustrating but results in an inferior mortar that does not gain full strength.

It is also a bad practice to set cement directly on a concrete floor such as in a garage or storage shed, as the dampness from the concrete will be drawn into the bag of cement. Most cement bags do come with a plastic liner inside, but moisture will still get into the bag if it is not closed tightly. When doing a job around the house, you don't want to waste your money by not properly storing the materials. If cement does partially harden in the bag, discard it; don't attempt to screen it and reuse it. If cement is stored correctly and kept dry, it will remain good for several years.

Cement will also have a tendency to cake or pack in a bag if left in one position for a length of time. If you know that you are not going to use your cement in a 60-day period, then it is a good idea to turn the bags over and change their position. All of the tips mentioned are the results of problems I have experienced over the years.

METAL REINFORCEMENT AND WALL TIES

There will be occasions when you will need metal reinforcement or wall ties to strengthen or tie the wall together. The most popular metal wire joint reinforcement for mortaring into the bed joints

Efflorescence—a whitish discoloration—is the result of excessive moisture that unites with a salt in the brick clay.

Efflorescence can be removed by wetting the wall and scrubbing it with a muriatic acid solution, but the stain may reappear until the bricks have dried out thoroughly. It's better to avoid the problem by keeping the bricks dry until you're ready to use them.

comes in 10′ lengths and resembles a track. The wire it is made of is about ¼″ in diameter, and it is available in different widths for varying wall sizes. The brand that I am familiar with is Dur-O-Wal, which is available from most building suppliers. Joint reinforcement is particularly valuable to reinforce masonry retaining walls of brick or block and to strengthen foundation walls where there is a lot of pressure from the earth.

There are a number of metal wall ties sold for masonry walls. The one you will most probably need is made of galvanized metal and is corrugated on the surface for better bonding in the mortar joints. These are commonly called brick veneer wall ties and are available at most building supply stores. They are approximately 8″ long and ⅞″ wide with two holes in one end. The ties are amazingly strong and flexible, allowing them to be bent as needed to fit into the mortar bed joints. They are sold for a few cents each or in boxes of 1,000.

Joint reinforcement or wall ties are laid in the mortar bed joints every 16″ in height (six courses of brick or two courses of concrete blocks). Wall ties are nailed into the wood studs about every 16″ on center, which is standard spacing for framing.

BONDS AND PATTERNS FOR BRICKWORK

Brick masonry can be built in a variety of patterns or designs, which are called bonds. Some are very simple, others are complex. Regardless of the type of pattern or bond that you select or that is shown in a project in the book, be sure to maintain the pattern throughout the wall or the end result will be disappointing.

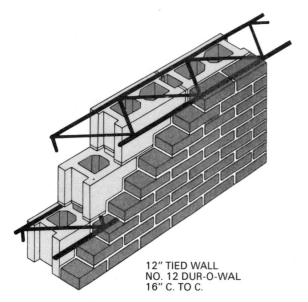

12″ TIED WALL
NO. 12 DUR-O-WAL
16″ C. TO C.

Metal wire joint reinforcement, shown here as used to tie together a wall of block and brick.

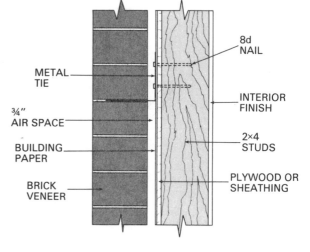

METAL TIE

¾″ AIR SPACE

BUILDING PAPER

BRICK VENEER

8d NAIL

INTERIOR FINISH

2×4 STUDS

PLYWOOD OR SHEATHING

Cross-sectional view of brick veneer tie mortared in and nailed to stud. A tie is used at every stud.

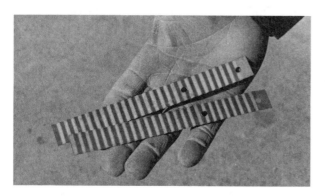

A common type of metal wall tie, called a brick veneer wall tie. The holes permit nailing the ties to the stud wall that backs the brick veneer.

To create these various bonds, bricks are laid in different positions. It is important that you know which position is referred to when you see it stated by name (see the illustration). A bond can be the same position repeated over and over again, or a combination of different positions.

Bonds in walls vary from bonds in paving work somewhat, although some bonds will work for both. The following bonds are the most often used for walls.

Half Lap Running Bond

This is the simplest of all bonds and is used more than any other arrangement. It consists of all full bricks (stretchers), each laid half over the one beneath it. The only cuts or half bricks used are at the corners or ends to start the bond. This bond pattern is also sometimes called the all-stretcher bond after the name of the position of the brick.

Common Bond with Full Header

The common bond, also known as the American bond, is much like the running bond but utilizes a course of header bricks to tie the wall together at regular intervals. The intervals are on the fifth, the sixth, or the seventh course, depending on whether the wall is backed up with brick or concrete blocks.

A 6″ cut piece of brick called a three-quarter is required on the end of each header course, either on or against the corner or end to create the 2″ lap over the brick below.

Common bond brick walls are very popular when a solid masonry wall needs to be built that is brick on one side and backed up with block on the other, as for a retaining wall, a garden wall, a garage, a garden shed, etc.

Common Bond with Flemish Header

A common bond brick wall can be varied on the header course by laying what is known as a Flemish

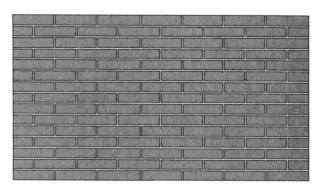

The half lap running bond, sometimes called the all-stretcher bond. It is the simplest and most common bond.

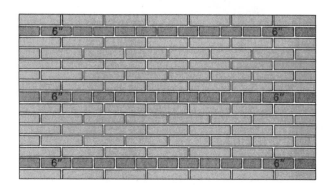

The common bond with full header, also called the American bond. Note the 6″ cut bricks necessary to create the proper 2″ lap.

Six positions in which bricks can be laid in a wall, with their names.

STRETCHER

HEADER

SOLDIER

SHINER

ROWLOCK

SAILOR

header. The only difference is that the header course is alternated with a header and stretcher brick. This arrangement is strictly only for appearance as both the full header and the Flemish header will tie the wall together strongly enough.

Flemish Bond

This bond pattern dates back to early times in England. It is considered one of the more beautiful bonds in brickwork. It is built by laying each course in alternating stretchers and headers with each header resting upon the middle of a stretcher below in succeeding courses. The headers in every other course are laid in a plumb vertical line.

A Flemish bond wall can be backed up with either bricks or blocks. If the mason is using block to back up a 4″ brick wall, however, the headers are cut into halves called snap headers. If snap headers are used, then metal wall ties or joint reinforcement must be used to bond the brick and block together.

There are two methods of starting the corners or ends in a Flemish bond wall: the dutch corner, in which a 6″ piece (three-quarter) is used, and the English corner, in which a 2″ piece (plug) is used. The Dutch corner is the more modern method; the English corner is the older traditional method. Purists insist on the English corner for a true Flemish bond, but you can take your choice.

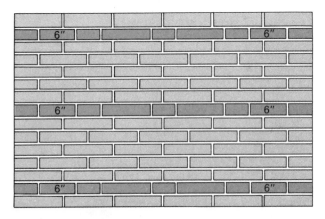

The common bond with Flemish header. In the header course, header and stretcher bricks alternate.

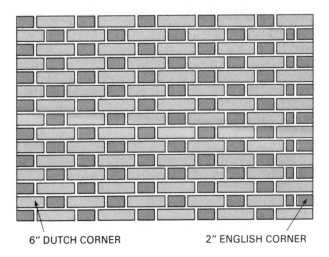

6″ DUTCH CORNER   2″ ENGLISH CORNER

The Flemish bond, showing a Dutch corner on the left and an English corner on the right. The English corner is older, but both are correct.

The Flemish bond wall has been made especially beautiful by raking out the mortar joints about ⅜″.

Double and Triple Stretcher Garden Wall Bonds

Flemish bond can be varied by increasing the number of stretchers between headers in every course. When there are two stretchers between each header, it is called a double stretcher garden wall bond. The design can be opened up by laying three stretchers between each header on the same course. This is a triple stretcher garden wall bond. The completed bond in a brick wall forms a repeating diamond pattern. Selecting dark bricks for the headers can create a further special effect.

Stack Bond

In a stack bond there is no overlapping of the bricks—all bricks line up vertically with those beneath. Although this bond is pleasing to the eye and is selected often by architects for ornamental brickwork, it is the weakest of all the brick bonds, since all the head joints are in a vertical line and are subject to cracking from expansion and settlement.

Metal wall ties and joint reinforcement are a must and are recommended every three courses. Bricks should be handpicked for length as much as possible to keep the vertical head joints lined up properly. This can be a problem, so consider this when making a decision on building a stack bond wall.

Paving Bonds

There are three primary paving bonds, the running bond, the basket weave, and the herringbone, but there are many combinations or variations of these. The running bond is the same as used in a brick wall. The herringbone is built of brick laid on a 45° angle from the border, with some cutting required along the edges. The herringbone bond will resist movements better than any of the others. The basket weave is simply an alternation of stretcher bricks, much like the weave of yarn or cloth.

All bricks vary slightly in length from burning in the kiln, and therefore slight adjustments will have

DOUBLE STRETCHER GARDEN WALL BOND
WITH UNITS IN DIAGONAL LINES

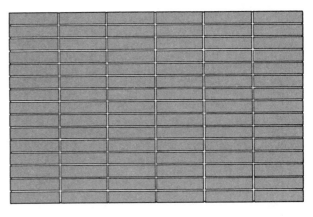

The stack bond—a very weak bond, for obvious reasons, but one often chosen for ornamental work because of its striking appearance.

TRIPLE STRETCHER GARDEN WALL BOND
WITH UNITS IN DOVETAIL FASHION

A double stretcher and a triple stretcher garden wall bond. These two variations of the Flemish bond require extremely careful bricklaying or the pattern will otherwise be glaringly distorted.

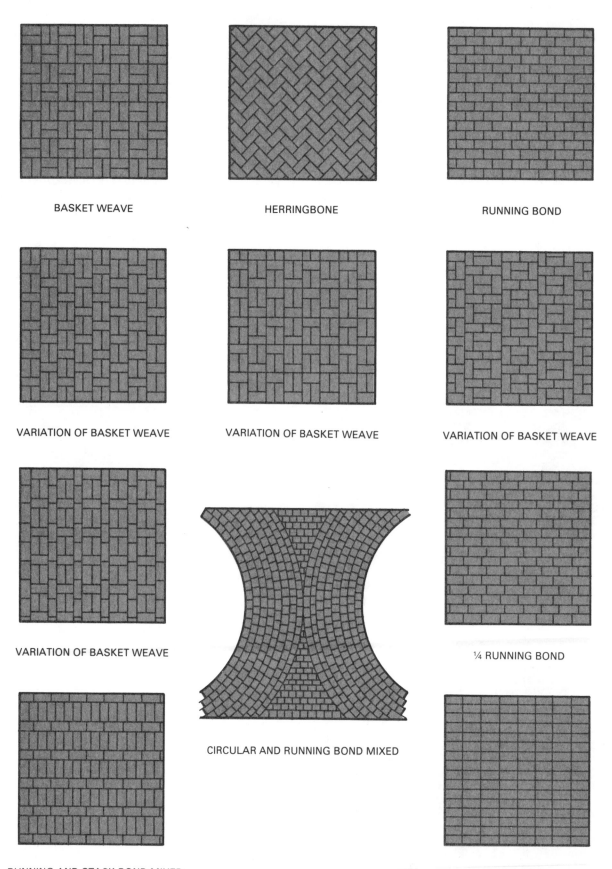

BASKET WEAVE

HERRINGBONE

RUNNING BOND

VARIATION OF BASKET WEAVE

VARIATION OF BASKET WEAVE

VARIATION OF BASKET WEAVE

VARIATION OF BASKET WEAVE

CIRCULAR AND RUNNING BOND MIXED

¼ RUNNING BOND

RUNNING AND STACK BOND MIXED

STACK BOND

A few of the many possible paving bonds. If you lay the bricks in a dry bed, you can try as many bonds as you like.

to be made as the work progresses. Paving bricks can be laid in a mortar bed or in dry sand or stone screenings, depending on your preference. There are various projects in the book that utilize both methods.

TOOLING OF MORTAR JOINTS

The treatment of mortar joints in the face of a wall affects the pattern and the finished wall texture. They provide a decorative effect on the surface of the wall by creating shadow or depth. At the same time, they have to provide a waterproof finish or the wall will leak.

Mortar joint finishes fall into two different groups: those made with the trowel and those formed with tools especially made for that purpose. The illustration here shows the different types of mortar joint finishes you can form with the trowel or a jointing tool. They include

- Concave or Round Joint—This joint is formed by running a rounded steel jointer through the joint and causes a slightly depressed, rounded joint finish. It is the most popular of all mortar joint finishes and is highly recommended for resisting rain penetration.
- V Joint—This joint is formed by running an

The herringbone paving bond is an excellent choice for broad areas because it resists movement better than the other bonds.

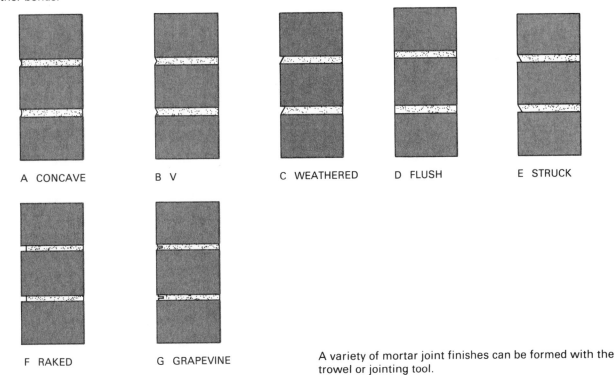

A CONCAVE   B V   C WEATHERED   D FLUSH   E STRUCK

F RAKED   G GRAPEVINE

A variety of mortar joint finishes can be formed with the trowel or jointing tool.

angled piece of steel attached to a handle through the joint. It forms a V impression in the joint and is especially recommended for a rough texture brick as it gives a sense of depth to the joint and creates a very attractive appearance.

- Weathered Joint—This joint is formed by running the trowel blade through the joints, holding it on an angle from the bottom edge and pressing the top edge in about ¼". It was used a lot in the past in older brickwork and is considered the best and most effective of all trowel joints.

- Flush Joint—This is the easiest joint to form and appears flat. It can be made by merely cutting the mortar off with the trowel flush and leaving as is. It is more effective and waterproof if the flat slicker tool is used, passing it very lightly over the mortar, smoothing the joint. Care should be taken to close up any holes tightly against all the edges of the bricks. It is highly recommended for any flat work such as paving, patios, window sills, etc.

- Struck Joint—This is a joint that was very popular in the past because of the speed at which it could be formed. The point of the trowel is held on an angle from the bottom to the top of the bed joint as it is run through the joints, pressing in the mortar about ¼" at the bottom and keeping it flush at the top edge. A small ledge is formed at the bottom and does not shed water as well as some of the other types mentioned. It is used today mainly when repairing or building an addition on an existing structure that utilized this joint.

- Raked Joint—This is a very attractive joint because it reveals the edge of the bricks and provides a dimensional appearance of depth. It is formed by raking out the joints with a skatewheel rakeout jointing tool that has an adjustable nail in the bottom. It is the most difficult of all the joint finishes to form correctly. Because it does rake out the mortar joint (¼" to ⅜" recommended), it is not as waterproof as the others, and special care should be taken to make sure that there are no holes left for water to penetrate. Any variation in joint size is also very noticeable in rakeout joints. Workmanship and quality of brickwork when a rakeout joint is used has to be of a higher quality than for normal brickwork.

- Grapevine Joint—This joint finish is created by a raised bead of steel on the tool. As it is passed through the mortar joint, it causes a varying indented line to appear. This joint finish was a favorite for early English and colonial brickwork and is still very popular throughout the United States.

TYPES OF BRICK WALLS

Solid Brick Wall

There are a variety of brick wall designs you can select for your project. The function, need, and requirements of the wall dictate the design selected.

The oldest type is a solid brick wall, which is faced with bricks and backed up by solid brick backing. Many garden walls are built this way. Brick header courses can be utilized, usually every 6th course, to bond the two tiers of brickwork together, or metal wall ties can be used to accomplish the same purpose, which would eliminate the header courses.

Brick and Block Wall

Brick walls backed up with concrete blocks are very popular for garages, foundations above grade, storage buildings, etc. The lower cost and paintability of concrete block make this a natural combination of masonry materials for an economical, strong wall. The brick and block can be bonded together every six courses of bricks high with metal wall ties, joint reinforcement, or by using brick headers. The fireplaces and chimneys in the project section are good examples of where brick and block work well together.

Brick Veneer Wall

Brick veneer walls consist of a brick face in front of some other material, usually wood framing such as found in the average house, townhouse, or garden apartment. Brick veneer provides an excellent combination of low-maintenance and energy conservation features at an economical price. Insulation is installed behind the brickwork and between the 2×4 wood framing. The nice part of brick veneer masonry is that the brickwork can be built anytime after the framing is erected without interfering with the interior completion of the project or structure. Weep holes are left at the bottom of the wall about every three bricks in length to drain any condensation that occurs inside the wall. Brick veneer can be built on a concrete block or poured concrete foundation that has been built below the freeze depth.

Brick veneer masonry is by far the most popular type of construction in the country today. If you are planning on an addition to your house, this would be the best method to use.

Brick Cavity Walls

Brick cavity walls are becoming more popular due to the high cost of energy in heating and cooling homes. This type of brick wall has a 4" thick exterior (single brick width) leaving a 3″ to 4″ airspace, then backing up with another 4″ brick wall or concrete block, whichever is preferred. The airspace in the center can be left open or insulation inserted as it is built for greater efficiency. Air will also insulate, but it is not as effective as insulations on the market today. This type of wall also requires that flashing be installed at the bottom of the wall and weep holes left about every three bricks in length to drain any moisture that condenses inside the wall or in the cavity. It has an excellent resistance to rain penetration and is considered to be the most effective type of brick wall construction. While it does cost more, it is worth the difference.

Reinforced Brick Walls

Reinforced brick walls are the strongest type of wall you can build of masonry, with the exception of solid stone walls. It is built like the cavity wall, only steel rods and concrete are placed in the center to make the wall stronger. The brickwork should always be allowed to cure or harden before attempting to pour the concrete in to prevent the

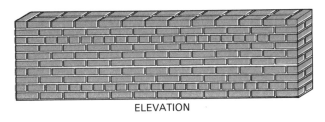

ELEVATION

A solid brick wall is one of a variety of designs available for your project.

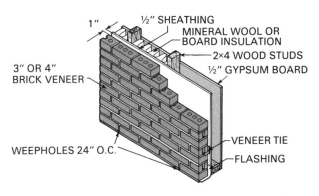

A typical brick veneer wall consists of a brick face in front of another material, which in this case is wood framing.

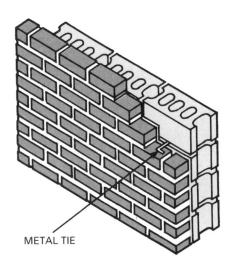

METAL TIE

A brick wall backed up with concrete blocks is another alternative. The brick and block can be bonded, as shown here, with metal ties.

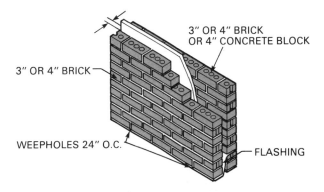

Brick cavity walls are becoming increasingly popular because of their greater effectiveness in heating and cooling. This advantage is credited to the nature of their construction as shown here.

walls from bowing out. Generally, the best practice is to build the brickwork up on both sides to scaffold height (about 5' high), let it cure at least 24 hours, and then pour the concrete inside the wall.

Projects such as retaining walls and swimming pool walls are excellent examples where reinforced brickwork would be utilized.

BUILDING BRICK MASONRY FOR SPECIAL PURPOSES

In addition to laying bricks to form walls, there are a variety of designs that can be built for special uses. The following examples are some of the more frequently used forms of brick masonry you may have occasion to build if designing your own project.

Piers or Columns

Piers are nothing more than brick posts or supports. They can be of different sizes depending on the requirements of your specific need. However, I would always try to make them utilize full bricks to simplify the job and present a more attractive workmanship appearance.

Piers are more difficult to build than regular walls because the construction involves mostly all plumb work and the slightest difference is noticeable in the finished job.

They are used in foundations for carrying or supporting beams and as supports under porches and decks, entrance posts at driveways, fence posts, etc.

Brick Wall with a Pilaster

When a pier is built as part of a brick wall and projects out from the face of the wall, it is known as a pilaster. These are frequently used in the construction of brick retaining walls for extra strength. If they only project on one side of the wall, they are called single pilasters. If they project on both sides of a wall, they are called double pilasters. As was the case with piers, try to make pilaster work full bricks with no cuts.

Brick pilasters are also used in the construction of garden walls to enhance their appearance, in addition to making the wall stronger. They should be interlocked or bonded into the brick wall with no stacked joints as shown in the following illustration.

Brick Chase

A brick chase is the reverse of a brick pilaster, with a recess being formed in the wall instead of a projection. The brickwork should also be laid out to work full bricks if possible with no cuts as shown in the illustration. The brickwork should be well bonded so that the wall will not be weakened at that point. Chases may be more than 4" deep if needed, but the wall will have to be built thicker to accommodate this.

Chases are designed so that electrical pipes, plumbing, and rain downspouts can be installed without projecting out past the face of the finished brick wall.

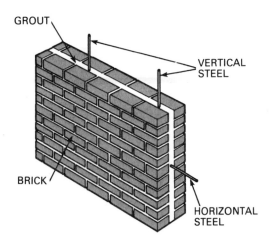

Components and construction of a reinforced brick wall make it the strongest masonry wall you can build.

Brick piers such as this are used as supports.

Brickwork Laid on Angle Iron Lintels over Openings

When brickwork has to be built over openings such as windows or doors, angle iron lintels are used to support the wall. If there is very little weight to be carried, pressed steel angle irons can be used; these are less expensive and are available at your building supplier. Generally, a heavier angle iron is used that is a full ⅜" in thickness and is available in 4" or 6" heights for heavier loads. A single 4" wall requires one angle iron; an 8" brick wall requires two 4" angle irons laid back to back; and a 12" wall requires three 4" angle irons with two being laid back to back and the other as shown in the illustrations. It is also important to set the front edge of any angle iron back from the edge of the wall about ½" so that mortar joints between the bricks do not pop out later. Angle iron lintels are available in a large number of lengths or can be cut to suit your needs.

Corbeling Out Brickwork

There will be occasions when you will want to rack out or corbel brickwork to make a wall of chimney wider at its top. A good rule of thumb

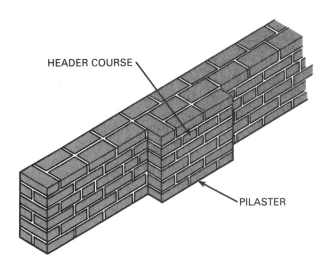

Brick pilasters should be bonded into the wall.

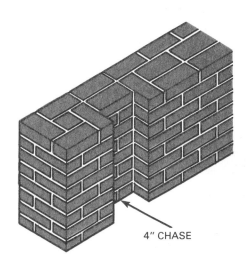

A brick chase is a recess formed in a wall.

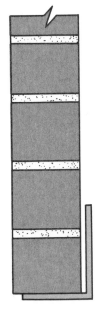

**4" WALL
ONE ANGLE LINTEL**

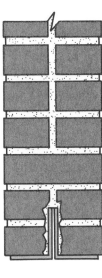

**8" WALL
TWO ANGLE LINTEL**

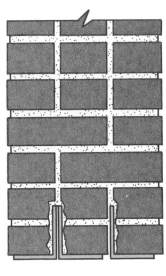

**12" WALL
THREE ANGLE LINTEL**

Use of angle irons with different size brick walls.

figure to use when corbeling is not to exceed 1" on any one course and make sure that the corbeling is bonded securely into the rest of the brickwork in the project. Form full mortar joints for maximum strength. The total projection of the corbeling should not extend more than one-half the thickness of the wall. If metal supports such as angle irons are used, this figure may be adjusted.

Fireplace construction requires a lot of corbeling inside the smoke chamber of the chimney to support the flue linings and reduce the total chimney size above the fireplace. This is shown in the project section with the fireplace plan.

Brick Windowsill

Brick windowsills can be laid in the rowlock or header position. They should project out from the face of the wall below about ¾" so the water will drip off. It is also imperative that full mortar joints be used for any brick sill work as there is a greater chance of leakage than in normal brickwork. Windowsills should be laid with a slope to the front so water will drain. Tool the mortar joints on top of the sill with a flat slicker joint to prevent moisture or water from penetrating. Generally, header position brick windowsills are installed when there are shorter window heights such as found in bathrooms.

MAKING A COURSE ROD FOR BRICKWORK

Before you actually lay any brick courses to form a wall or project, the coursing heights must be worked out. You can measure with the mason's scale rules and check each individual course to see that it is correct, or you can make a wooden course rod by marking on it each course of bricks as they should be laid. The course rod would be the most efficient method, and you would be less likely to make any mistakes. Courses of bricks marked on the rod always include the mortar joints, so you need to make no allowances for them. The plans in the project sections indicate the proper coursing; however, there may be times when you will want to figure out your own. To understand how to lay out a course rod, you need to know how to read and interpret the two types of masonry scale rules in use. An explanation is offered for each, with the differences in the scales noted.

The following illustrations show the two types of scale rules used for laying out brickwork for height.

The Spacing Rule

The spacing rule, shown at the top of the photograph, was designed long before the modular system of building was developed. By modular, I mean

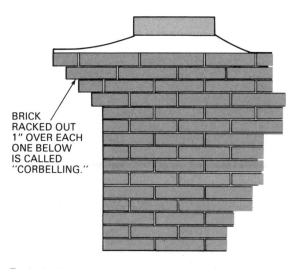

Typical chimney corbeling of cap.

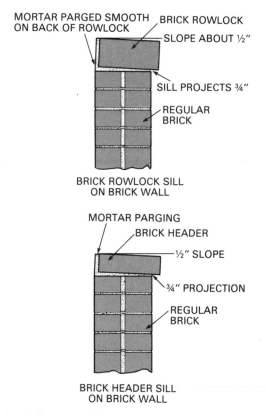

Sectional views of brick rowlock and brick header sills.

masonry materials such as brick, blocks, tiles, etc., which are manufactured on a 4″ grid. This is done to have building materials work out with a minimum of cutting. The spacing rule was designed for brickwork to divide mortar joints evenly in a wall when there was a measurement that did not come out modular or in multiples of 4″.

Let's use an example to see how this applies. You are building a window in a wall and the height of the window is 49½″. Since the modular rule will only work with multiples of 4″, it could not be used to mark off a course rod in equal courses. The closest measurement that would work for the modular rule would be 48″. As you can see, there is an additional difference of 1½″ for which to compensate; therefore, the spacing rule will be used.

Looking at the standard side of the rule, turn it over slowly and notice that 49½″ lines up with number 6 on the scale side. (This number 6 does not mean the same as on the modular rule.) Using this scale, all of the mortar joints will be slightly thicker to make up the 1½″ required. This will then work out perfectly with the 49½″ measurement and solves the problem. Each mortar bed joint is increased the same exact amount so that even though all mortar bed joints are slightly thicker, they will not be noticeable when one looks at them.

Mortar bed joints can also be decreased, making them thinner, if necessary. There is, however, a limit as to how thick or thin of a mortar bed joint can be used and still build a brick wall, making it presentable. My recommendation is to not go below number 4 for thin joints or above number 6 for thicker joints.

Another important feature of the spacing rule is that the total number of courses shown on the scale side from the bottom to the top of the rule is marked in red ink. This is handy when you want to estimate the number of bricks needed for a wall or project.

The height of a wall can be checked; at the same time this measurement determines the number of courses needed. For example, a brick project that is 60½″ high will take 22 courses of bricks laid on number 6 on the spacing rule.

Spacing rules are used frequently in building brick windowsills, as window frames are seldom modular in length.

Modular Rule

The modular rule, shown at the bottom of the photograph, also has the standard 72″ on one side and a scale on the opposite side. As mentioned, the scales on the modular rule represent masonry units

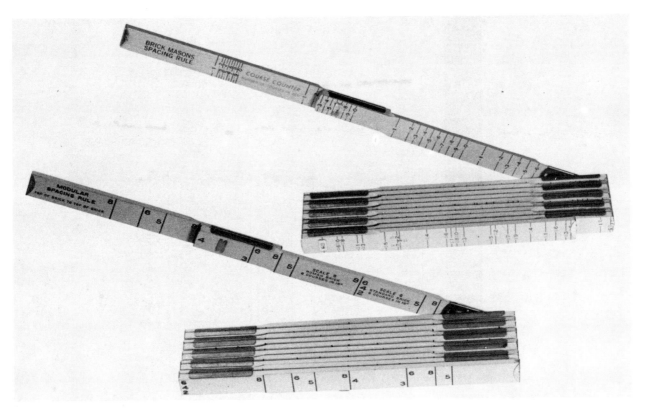

Masonry scale rules are available in these two forms.

that are made to conform to the modular system such as standard bricks, oversize bricks, tile, concrete blocks, etc. Normally, most brickwork is built of standard bricks with a ⅜" mortar bed joint. This is number 6 on the modular scale, because there are six courses to every 16". At multiples of 16" on the regular side of the modular rule, it will be marked in red ink to remind the person who is using the rule that different kinds of masonry materials will bond together at this height. The same is true for steel tapes used by the carpenter. They have a diamond mark every 16" to indicate the modular layout of wood framing.

To state another example, six courses of brick are laid to number 6 on the modular rule. They are backed up with two courses of 8" concrete blocks. Since each concrete block is 8" high including the mortar joint, it takes two courses to become even with the brickwork; therefore, scale 2 is used for concrete block on the modular scale rule. This information will come in handy when you work on block projects.

To simplify reading the modular rule, most of them have a printed notation right on the rule next to the number on the scale.

The following two illustrations show how the two scales are used with brickwork.

Laying Out a Course Rod for a Brick Wall

The illustration here of the front elevation of a brick garden shed will be used as an example for laying out a typical course rod. Since the wall measurements are laid out in multiples of 4", the modular rule will be used.

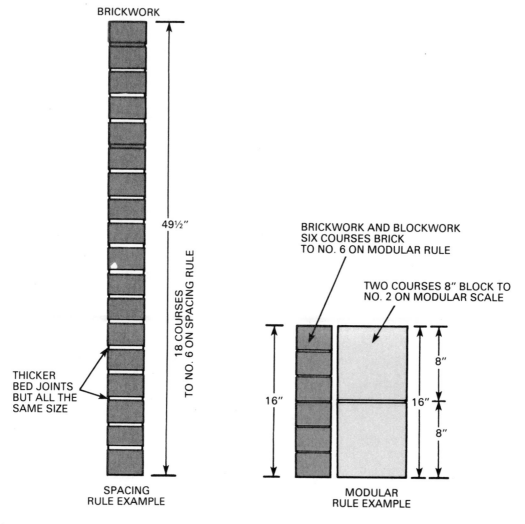

The spacing rule and the modular rule can be used as shown here.

1. Select a straight wood rod such as a 1×3 furring strip. Referring to the illustration, notice that the wall is 88" high. Checking this out on the modular scale, we find that this works out to perfect courses on number 6 and is 33 courses of bricks high. Square a line across at the 88" height and cut the rod off with a hand saw. Mark the top with a crowfoot (arrow) and print "top" so it does not get used upside down. Using the rule, measure up from the bottom of

the rod and mark the height of the window and door head. As you can see on the plan, they are both at the same height. This measurement is 6'8" and 30 courses of bricks high.

2. Next, measure up from the bottom of the rod and using the square and pencil, mark the top of the brick windowsill, which is 32" and 12 courses of bricks high. Since this is a brick header sill, it will work a regular course height. If it had been a rowlock brick position, a 4"

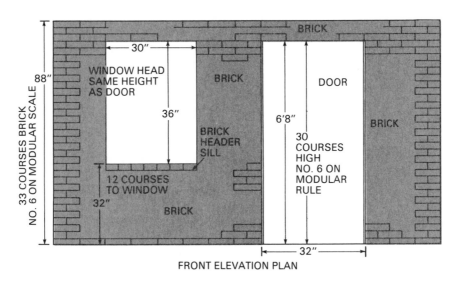

Brick garden shed example for laying out a typical course rod.

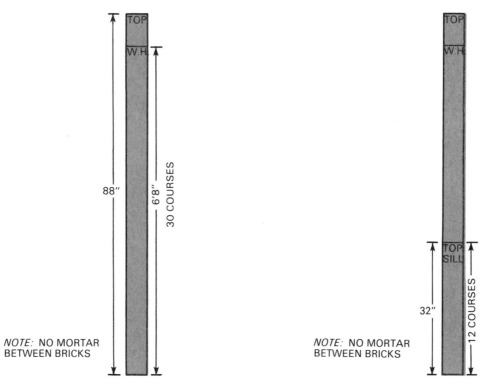

Course rod with window and door head marked.

Course rod with sill marked.

allowance would have been necessary. When this is completed, print "sill" to remind you of the exact height when building the wall.

3. Now, lay out from the bottom to the top, the individual courses of bricks by holding the modular rule at the bottom and marking a pencil point at number 6 on the scale. Using the square, extend each of these lines across the rod. Since the rod is laid on the modular scale, you will find that the coursing will work even with all of the measurements.

4. Last, number from the bottom to the top of the rod, indicating each course of brick in the wall. Recheck all measurements when this is done to make sure you have not made any mistakes.

After the course rod has been checked for accuracy, the coursing marks can be made permanent by sawing in with a hand saw about ⅛" deep. This will prevent the marks from fading out until the project is finished.

Shorter versions of a course rod can be made to fit any project you build. The rod is more accurate and a lot less trouble to use than constantly unfolding and folding a rule. On a project over 6' high, which is the length of a regular mason's folding rule, I would strongly recommend making a course rod, as there is too much chance for error when sliding the rule up past the 6' height.

When using the course rod on your project, remember it must be set at the same level base point on all corners if it is to be correct. You can establish these bench marks, as called in the trade, by using a long straight edge board with a level laid on top or a garden hose with water in it.

Take care of the course rod by keeping it dry when not in use. This will prevent bowing or warping and ensure it will be in good condition when you need it.

BASIC SKILLS OF LAYING BRICKS

The basic skills of laying bricks look simple when you watch a professional bricklayer work. Such skills are, however, developed by practice—doing the same thing over and over again. The difference between the professional mason and you is that it may take you longer, but if you study and imitate the movements and techniques shown, your work can be just as good. It is going to take time and patience, but you shouldn't begrudge either if you expect to obtain the best results.

The close-up photographs and descriptions of these skills and techniques show the little things and methods that are hard to explain in words only—holding a brick properly, cutting off the mortar with a trowel so as not to smear the wall, leveling, plumbing, and striking (tooling) of the mortar joints.

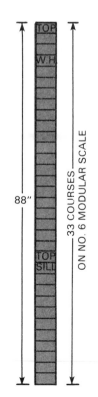

Course rod with individual courses marked.

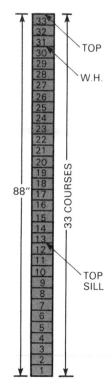

Completed course rod with brick courses numbered.

The photographs and comments are based on my experience of working in the masonry trade all of my life and teaching bricklaying for 17 years. In this period of time, I have attempted to analyze each step of laying bricks, and I believe that after reading this section you can acquire the necessary skills to build your masonry project.

Laying Out the Corner or Wall

All brick projects are going to require striking a line and laying out the work before actually laying any bricks. This is done to eliminate as many mistakes as possible. Usually a corner is going to be required, if the wall or project has a return.

The first part of this section, therefore, deals with the laying out and building of a brick corner. Any blockwork will be covered in the chapter on concrete block.

The most common corner is the right angle. To determine how many bricks will be needed for the first course, remember this rule: The number of bricks on the first course will equal the number of courses in the completed corner. This is because a corner in standard running bond (half lap) will rack back half a brick on each side per course, and therefore a brick corner nine courses high will require a first-course layout of five bricks in one direction and four bricks in the other direction.

To make sure that the mortar will adhere to the concrete, be sure to brush off the footing or base. Then strike a chalk line to the angle of a square.

Next, dry-bond (lay out without mortar) the bricks out to the lines to establish the bond. This is done by inserting the tip of the little finger between each two bricks. Don't put your finger all the way down in the joint; just use the tip to make a space that in most cases should be about ⅜".

Lay out the best face or side of the bricks to the line as they will be laid in the mortar. Almost all bricks have one good straight side; the other side

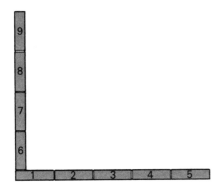

Layout for the first course of a brick corner to be nine courses high.

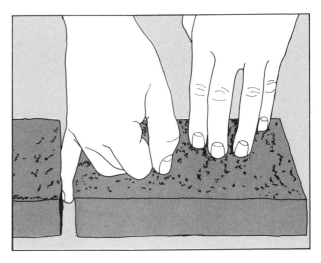

When dry-bonding the bricks, use your little finger as shown to maintain about ⅜" between bricks.

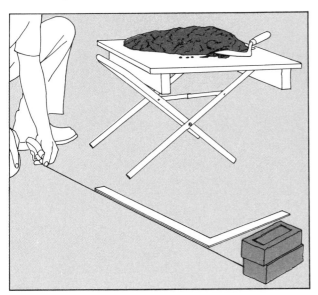

A carpenter's square will give you a good right angle. Strike a chalk line for the first course.

usually has a wrinkle or slight bow, because bricks lie on their sides in the brick kiln when being burned. Mark the end of every other brick with a pencil on the concrete.

Cutting and Spreading the Mortar

Mix the mortar to a stiffness such that it can be spread with the trowel and the bricks can be pressed into the mortar without excessive tapping on them with the trowel handle or blade. You will soon learn this consistency after laying a few bricks.

A good simple test to determine if your mortar is rich or sticky enough is to pick up some on the trowel, set it on the blade with a downward jerk of the wrist, and then turn the trowel upside down. The mortar should stick to the trowel. If it doesn't, it may not stick to the bricks either.

The simplest method of cutting mortar from the board is by "cupping." This is done in four different movements. Practice them until you can spread mortar for at least two bricks in length. Cupping is not difficult to master, and it is a very important skill.

Pick up and hold the trowel so that your thumb is about even with and slightly over the front section of the handle. Grasp the trowel firmly but not so tightly that your fingers cramp. Then cut, shape, pick up, and set the mortar on the trowel as shown in the illustrations.

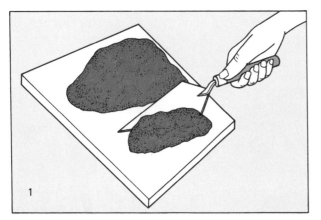

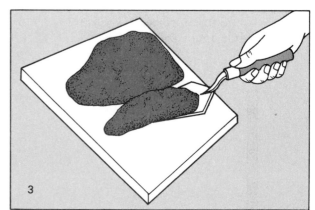

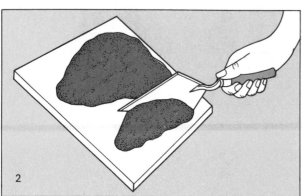

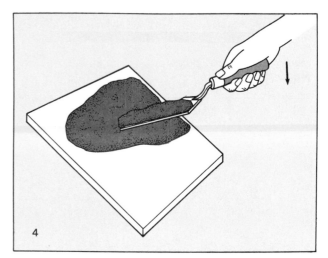

The four-movement process of cupping the mortar from the mortarboard. (1) Cut down into the mortar and pull a trowelful free. (2) Shape the mortar to fit the trowel by drawing the blade against the sides of the lump. (3) Quickly slide the trowel under the mortar. (4) Pick up the trowel, then snap your wrist downward to set the mortar on the blade. Now it's ready to spread.

Spread the mortar by using a sweeping motion, trying to keep the point of the trowel in the approximate center of the wall line. The mortar should roll off the trowel smoothly. Practice is the only way to learn to perform this effectively.

Complete the mortar bed by furrowing it with the point of the trowel. Don't press so hard that the point of the trowel goes all the way through the mortar, as this could cause the joint to leak.

Laying the Bricks

If you're standing on the back side of the wall, pick up the brick to be laid, with your fingers over the front edge and your thumb to the back. If you're working from the front of the wall, then your thumb should be in the front and your fingers in the back. This is important to master in the beginning, because if you hold the brick the wrong way, when you're laying bricks to the line your

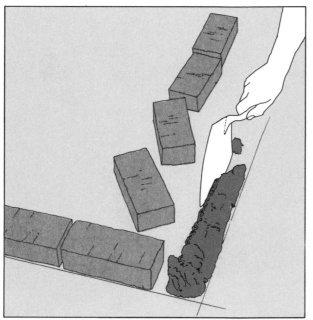

How to hold a brick to be laid when you're working from the back side of the wall.

Spread the mortar along the center of the wall line. Don't be too forceful or jerky or the mortar will not spread smoothly. Then furrow the mortar bed by pressing with the point of the trowel in a hopping motion. Don't go all the way through the mortar.

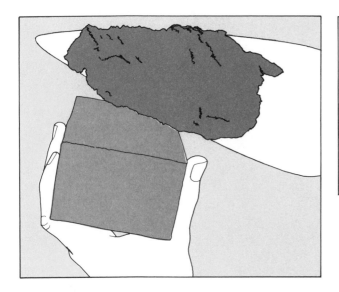

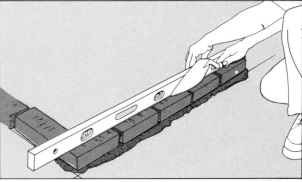

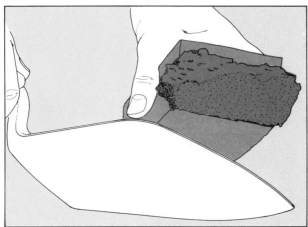

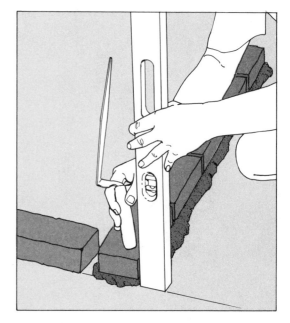

Apply mortar to the end of a brick about to be laid by holding the brick at an angle and swiping on the mortar in a downward motion.

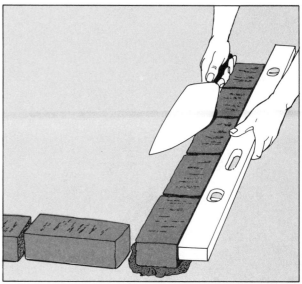

Leveling, plumbing, and aligning. When tapping bricks with the blade or handle of the trowel, be careful not to tap on the outside edge of a brick, or it may chip.

fingers will keep moving the line, making it difficult to build a straight wall.

The mortar head joints (on the ends of the brick) can be applied either by swiping the mortar against the end of the brick in place or by applying it to the brick about to be laid. In my opinion, it is easier to apply it to the end of the brick to be laid.

Do this by holding the brick with the end facing you on a slight tilt. Pick up a small amount of mortar on the front section of the blade and swipe it on the end with a downward motion. With some practice, this will fill out the end of the brick fully.

Once the first course of bricks are laid in position, level them by laying the level in the center of the course and tapping down any high bricks. All of the bricks should be against the edge of the level, with no high or low places. The bubble should be between the two lines on the vial. Any low bricks should be taken up and relaid with fresh mortar.

Plumb each end of the course, making sure that the plumb bubble is in between the lines on the vial and the level is against the face of the brick.

Align the entire course by placing the level at the top outside edge of the bricks. Adjust the bricks so they are in a line against the level. Many bricks have a curled edge at the top to simulate old handmade colonial bricks. Line up with this curled edge. This is important if the wall is to be straight and plumb.

Complete the first course of bricks by cutting off the excess mortar at the bottom of the mortar joint and pointing any holes with the point of the trowel.

Check the height of each subsequent course by measuring with the mason's modular scale rule.

Plumbing is the most difficult phase of laying a brick project. It can be made a lot easier if you position your head directly over the corner or edge of the brick when laying them and sight down the wall.

Trim the mortar bed under the first course of bricks with the trowel, filling in any holes.

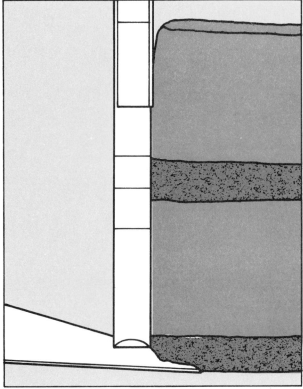

The simplest way to check course height is to rest the trowel tip on the brick below and set the rule on it. The number 6 should align with the top brick.

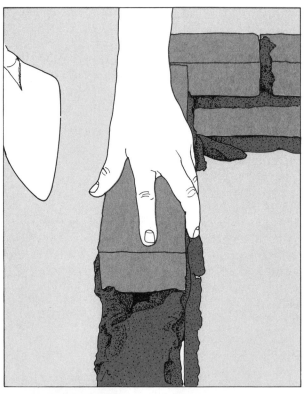

Plumbing bricks by using the ''sighting'' method, looking straight down at the corner. This is an important bricklaying skill that will prove to be valuable.

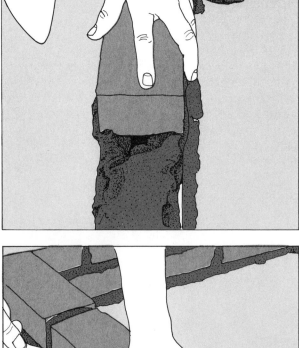

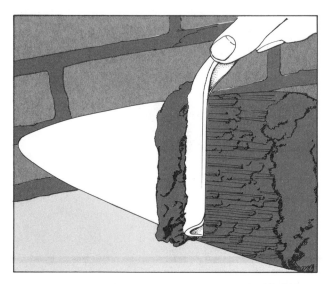

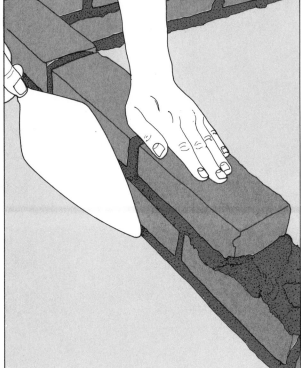

The position of your hands is important. As you press the brick into position, rest your palm both on it and on the brick previously laid, so that you can feel when the two are approximately even.

Pick up mortar with the striking tool with a slight pinching, rolling motion against the trowel to make the mortar stick to the tool.

Tooling (Striking) the Mortar Joints

As soon as the mortar joints have set (dried) enough to smear, they should be tooled. Not all of the mortar joints are going to be completely full, so it will be necessary to point any holes as you tool them. Press your thumb into the mortar joint. If it leaves an impression, the joint is ready to be struck. Never let the mortar joints get too hard before striking or they will turn black when in contact with the metal striking tool. This is known as "burning the joints" and can ruin the appearance of your brick project. Timing is very important in tooling the joints.

Always tool the head joints first, pressing the tool firmly into the joint. It should smooth and fill out the mortar against the edges if done correctly. Then tool the bed joints last so that you have a straight continuous line when looking at the joints.

After the mortar joints have dried enough not to smear, brush them lightly. If necessary, restrike to remove any particles of mortar and give the joint a neat appearance.

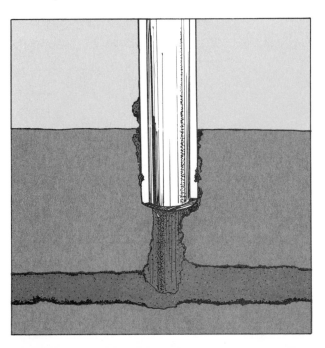

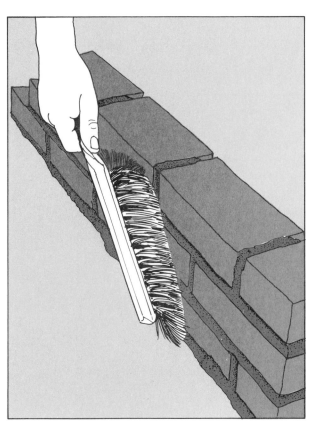

Brush the mortar joints after they are dry enough not to smear, to remove particles of mortar.

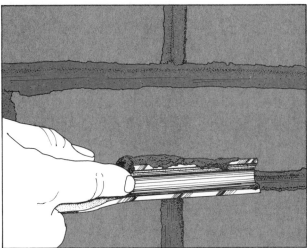

First tool the head joints, then the bed joints, drawing the tool firmly in a continuous horizontal line.

Complete the corner by lining up the ends of each brick. This is done by holding the level against the point or end of each brick from top to bottom and adjusting them to the level. Restrike mortar joints that were "cracked" when tailing the corner.

Laying Bricks to the Line

The most accurate method of building a brick wall that is over the length of a level (4') is by attaching a line to each end or corner and laying the bricks to the line.

I strongly recommend buying nylon line, as cotton will not hold up and breaks easily when stretched tightly. As mentioned in the suggested tools and equipment, select a braided nylon line, about #18 in diameter.

Before laying any bricks to the line, follow the techniques previously discussed by dry-bonding the first course and building a corner or end on the project.

There are two methods of attaching a line to the corner or end of the project. The first is by using wood or plastic line blocks. Line blocks can usually be obtained from your local masonry material supplier for the asking. They are held to the wall by tension, being stretched between and hooked on the ends.

If it is not practical to use line blocks, a nail and line pin can be used in their place. Line pins can also be obtained from building suppliers for the asking.

To lay a brick correctly to the line, it should be flush with the face of the brick below and the top

"Tailing" a corner—that is, lining up the ends of each brick from top to bottom. This is important if a straight wall is to be joined to a corner.

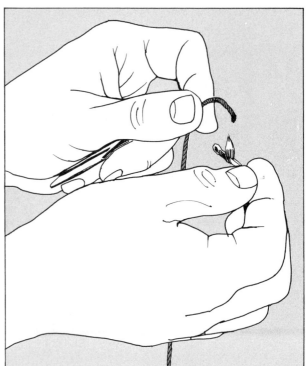

Braided nylon line is best to use as a line. Heat the ends with a match to keep them from unraveling.

edge should be about $^1/_{16}''$ back from the line. In addition, the top of the brick should be even with the top of the line. This $^1/_{16}''$ measurement is not measured with the rule but is an approximate distance. You should see a little light between the edge of the brick and the line. If bricks are allowed to touch and push the line, the wall will be bowed out. This is known in the trade as "being hard to the line." If the bricks are back too far from the line, then a hollow place will develop in the wall. This is known as "being slack to the line." Both conditions are bad, so be sure to position your brick the correct distance from the line.

If you are going to build a long wall (in excess of 30'), it may be necessary to lay a brick in the center of the wall for plumb, level, and the proper course height, and attach a metal trig to prevent the line from sagging. The trig is a simple metal gadget that can be clipped on the line and then set on top of the trig brick. A piece of brick is laid on top of this to hold it in position. On windy days, the setting of a trig is a necessity to keep a straight wall. Metal trigs

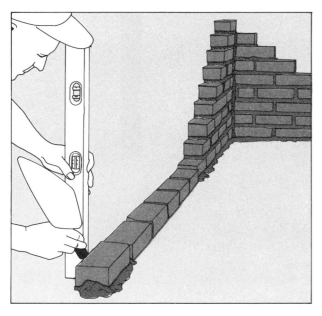

This wall will be long enough to benefit from the line technique. The first course is dry-bonded, with the end brick set in mortar and plumbed.

Laying bricks to line. Here line blocks are being used.

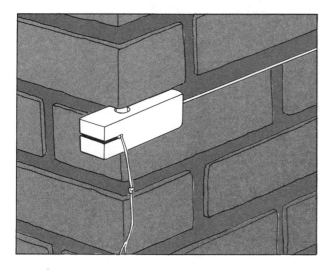

Attaching line with line blocks. Tie a knot in one end of the line and slip the line into one of the blocks. Stretch the line to the other end of the wall, slip it into the other block, and wrap it around so that it won't slip. Tension will hold the blocks in place.

are given away as a form of advertisement by building suppliers who sell masonry materials. A piece of string looped and tied around the line will do the same thing if you cannot obtain a free one.

Laying the Last Brick in the Wall

It is very important that the last brick laid in the wall have well-filled mortar joints, because this is the place a wall is most likely to leak. To achieve this, butter (apply mortar) to the ends of both bricks adjoining it and then butter both ends of the brick being laid. When tooled, this will form a full head joint that will not leak. The last brick laid to complete the course is known as the closure brick.

Cutting Brick

Many times in the course of your work, it will become necessary to cut a brick. They should be cut accurately for a good appearance. One method is to use a brick hammer. Don't try to break the brick completely through until it has been scored all the way around. Then turn it in your hand back to the face side and give it a sharp blow with the blade of the hammer. It should break cleanly with any luck. If not, score it again.

A neat, more accurate cut can be made with the brick set chisel, which was described in the list of recommended tools. Place the chisel on the line with the flat side toward the piece you want to use.

Using a trig. First a brick is laid in mortar in the middle of the course, with a loose brick on top to hold the metal trig. Then bricks are laid to the corners from each side of trig.

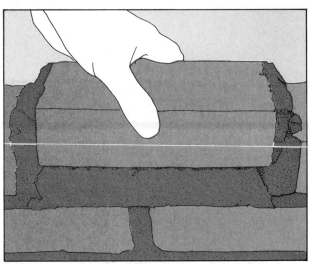

When laying the last brick in a wall, apply mortar to the ends of both the bricks being laid and the two bricks on either side to ensure well-filled joints.

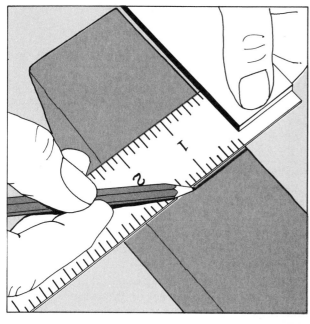

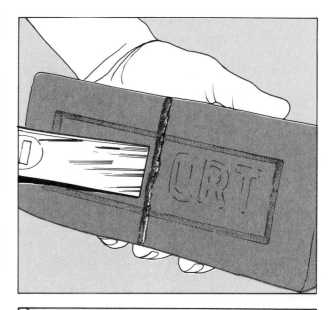

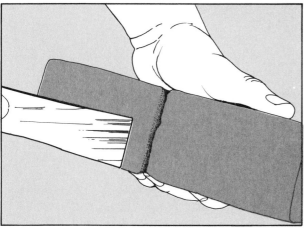

Cutting brick. First mark the desired cut with a small square. Score the face along the line with the hammer blade, then continue scoring all the way around.

Strike a sharp blow with the hammer on the head of the chisel. It may take several blows to break the brick. Trim off any rounded edges on the back side of brick that did not cut cleanly.

CLEANING BRICKWORK

Your brick project can be greatly improved and its full color brought out by cleaning it. The cleaning should not be done until it has cured for at least one week so the mortar joints are not damaged by the cleaner.

Some extra care taken when building the project will pay dividends when it is time to do the cleaning. Practices such as keeping the mortarboard back from the work to prevent splashing mortar, smearing the face of the bricks excessively by holding the trowel on too flat of an angle when cutting the mortar joints off, tooling the mortar joints when they are too wet and brushing the brickwork before it has dried enough are some of the most frequent causes of messy brickwork.

Protect your clothes during the cleaning process by obtaining a rubberized or plastic pair of pants such as commonly found in sporting goods stores or by wearing rubber gloves, rubber overshoes and an old shirt or jacket, and, of course, some type of hat to protect your head. Safety goggles or glasses will protect your eyes but do present a minor problem in that they will fog up or become wet from the washing process.

The cleaning tools are few and inexpensive. Most of them you probably have laying around the house now: a metal scraper, stiff bristle brush such as used in dairy houses, pointing trowel, plastic bucket or pail, garden hose with a nozzle attached, and a couple pieces of bricks to rub stubborn spots.

There are a variety of chemical cleaners available from building suppliers for different purposes. A number of these will be discussed in the maintenance and repair section of the book later. The most often used traditional cleaner is muriatic acid diluted with water. This is readily available from

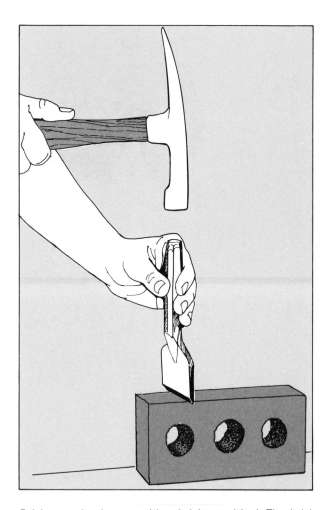

Brick can also be cut with a brick set chisel. The brick should be cushioned on a piece of wood, and the flat side of the chisel should be toward the piece you want to use. Don't hit your hand.

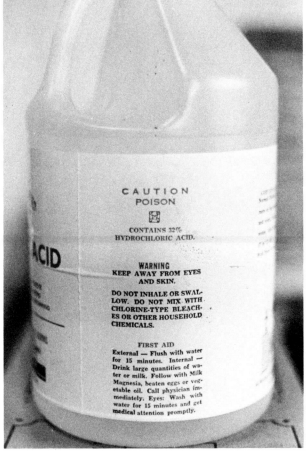

Muriatic acid is the most often used traditional cleaner.

almost all building suppliers or hardware stores. It is packaged in quarts or gallons in plastic containers. Be sure to read all directions carefully on the bottle or label before using.

I have used the following procedure for many years with a lot of success when cleaning new brickwork. It is called a "bucket and brush" method.

1. Start by dry-cleaning the brickwork by scraping off any loose mortar particles with the scraper or pointing trowel. Use a chisel for any large chunks, being careful not to ruin the face of the bricks. After this is completed, brush the brickwork with a good stiff-bristle brush to remove all of the loose dirt particles.

Step 1

Step 2

Step 3

Step 4

Step 5

Step 6

2. If there are any metal windows, doors, etc., it would be a good idea to tape some plastic covering around them to protect them. Also, any shrubbery or flowers should be covered with a sheet of plastic as the cleaner could "burn" them. It is better to be safe than sorry later!

3. Add about 1 gallon water to the plastic bucket first, from the hose.

4. Pour the acid into the water carefully, in an approximate amount of 1 part acid to 10 parts water. You can guess at this measurement as it is not that critical, or use a plastic measuring container if desired. This is a fairly weak solution and will not damage the brickwork. Later, if you need a stronger solution due to a lot of hard-to-remove stains or smears, add a little more acid to the water.

   I strongly recommend that you wear safety eye protection when mixing the acid solution and avoid inhaling the fumes by not standing directly over the bucket. To determine if the acid solution is strong enough, dip your brush into the bucket and flick a small amount on the brick surface to be cleaned. It should foam on the surface.

5. Before applying any acid solution to the brick surface, soak the entire area with water from the hose with a nozzle attached. Start at the top, using a spray, and saturate the wall until the water runs down the surface.

6. Starting at the top of the project (brickwork), scrub vigorously with the brush, dipping into the acid solution and working your way down to the bottom. Don't scrub too large an area that the cleaner starts to dry or it may become difficult to rinse it off with water. A good suggestion would be to try an area 4' wide by 4' high. The right amount of area would be determined by how the drying conditions are at that particular time. As soon as you notice the wall drying out too fast, start rinsing well immediately with water.

7. When encountering stubborn stains, smears, or troublesome spots of mortar, rub them with a piece of brick and rescrub until clean. If any of the acid solution does splash on you, rinse off immediately with running water from the hose. Muriatic acid solution will dilute easily with water.

8. As soon as you are satisfied that the brickwork is clean, flush the wall well with water from the hose until it runs clear. This will also neutralize any acid solution left on the surface. When the cleaning is completed, be sure to rinse any of the tools or rubber clothing used with water to prevent them from being damaged.

32
CONCRETE BLOCK TECHNIQUES

This chapter contains information on popular types of concrete blocks available, the different sizes you can obtain, where to buy them at the best price, how to estimate concrete blocks and mortar for a job or project, and how to start the corner for different block widths. There are only a couple of patterns or bonds to use with block. These are the same as described in the brick section.

In addition, there is a complete illustrated description of how to lay concrete blocks. The principal differences between bricks and concrete blocks are in the design of the block and its weight and size. Techniques of laying blocks, spreading mortar, cutting, etc., are a little different from those for laying bricks. As you build the projects in the book, refer to this section as often as necessary for tips and techniques.

Let's begin with a little background on blocks. They are not a new product, but have been around for about half a century. They are called concrete blocks because they are made of portland cement and some type of aggregate (course filler) such as cinders, stone screenings, or lightweight materials like expanded shale or fly ash. Concrete blocks are not burned in a kiln like bricks, but cured or hardened with steam. It takes only about four hours to season blocks. Because of this, there is very little shrinkage or expansion; you will find them very accurate in dimension.

Blocks are popular among builders because they are durable, economical, strong, fire-resistant, and easy to lay. They go up faster than bricks, because of their larger size. It would take 12 bricks to fill the same area as one 8"×8"×16" block. They can be combined with bricks or stone to form an economical wall. For this reason, they have been a choice for foundations for many years.

In addition, concrete blocks are available in all parts of the country, in the same standard sizes and types to suit any job requirement. Blocks can be painted to serve as a finished wall, and they have good insulation value because of the air space inside the hollow cells. In recent years, there has been a trend to place insulation in the cells of the blocks to provide extra insulation.

For years concrete blocks were made with three cells (holes) in them. Now many manufacturers are also making blocks with two cells in order that the above-mentioned insulation or poured concrete can be added more effectively. They are mostly made square on both ends, so you don't have to worry about ordering extra corner blocks.

Until they are laid in the wall, blocks are fairly delicate. By delicate, I mean that they cannot be dropped or slammed on a flat surface or they may break. Be especially careful not to drop a block on your feet.

TYPES OF CONCRETE BLOCKS

There are a large number of different types of concrete blocks you can select from, depending on your project. The illustration shows some of the more popular ones.

Concrete blocks have good insulation value because of their air spaces, and it can be increased by inserting insulation in the spaces.

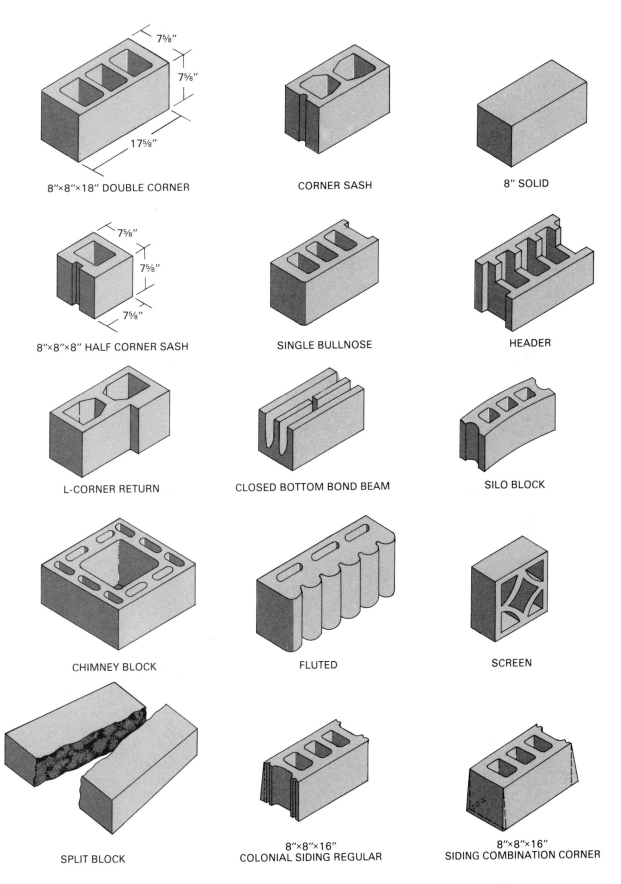

8"×8"×18" DOUBLE CORNER

CORNER SASH

8" SOLID

8"×8"×8" HALF CORNER SASH

SINGLE BULLNOSE

HEADER

L-CORNER RETURN

CLOSED BOTTOM BOND BEAM

SILO BLOCK

CHIMNEY BLOCK

FLUTED

SCREEN

SPLIT BLOCK

8"×8"×16"
COLONIAL SIDING REGULAR

8"×8"×16"
SIDING COMBINATION CORNER

Some of the many types of block available.

Block Sizes

Blocks, like bricks, are made modular (to a 4" grid) so that they will work out with bricks, stone, and other masonry materials. Standard concrete blocks are always the same length and height, but come in various widths for different wall thicknesses. Blocks are made to be laid with ⅜" mortar head and bed joints. Therefore, the actual measurements of a block are ⅜" less than the expressed figure for a finished block. For example, the standard concrete block is considered to be an 8" block that measures 15⅝"×7⅝"×7⅝". After the mortar joints are applied to the block, it rounds off to 16"×8"×7⅝". This is known as the nominal size. Whenever you order blocks, always state the full size including the joints. It would be ridiculous to pick up the phone and ask for a 15⅝"×7⅝"×7⅝" block when one could merely ask for so many 8" blocks. The length and height are taken for granted.

The stress the wall will be subjected to, or where it is going to be, will determine the width of block to buy. Blocks range in width from a 2" slab used for patios or paving to 12" blocks used in foundation or retaining walls. For example, a foundation wall should be built of either 10" or 12" concrete blocks. However, a block partition inside a building that carries no heavy load can be either 4", 6", or 8". The 6" block is a good compromise for inside partition walls.

Specialty Blocks

There are a large number of special blocks you can select from, depending on your project and the effect you want to achieve. Screen blocks are used a lot to provide privacy around patios, pools, etc. They are ornamental and their open areas allow cooling breezes and light to pass through. Fluted blocks have become very popular in the last five years. Fluted blocks have projecting vertical ribs, which create a textured appearance, and are usually laid in a stacked bond pattern. Another decorative block, known as "splitblock" or "splitrock," is made by splitting a solid 4" concrete block in half lengthwise; splitblocks are laid in the wall with the rough face showing. Usually a white or gray aggregate is used to achieve a stone appearance.

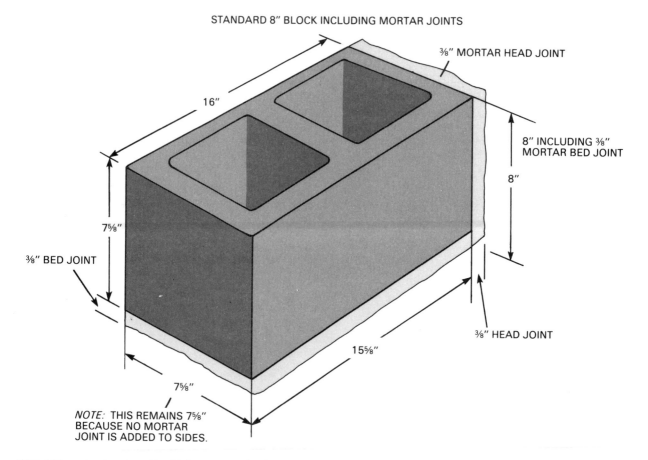

STANDARD 8" BLOCK INCLUDING MORTAR JOINTS

⅜" MORTAR HEAD JOINT

16"

8" INCLUDING ⅜" MORTAR BED JOINT

8"

7⅝"

⅜" BED JOINT

⅜" HEAD JOINT

15⅝"

7⅝"

NOTE: THIS REMAINS 7⅝" BECAUSE NO MORTAR JOINT IS ADDED TO SIDES.

With ⅜" mortar head and bed joints, a standard concrete block fills an area 8" high and 16" long.

BUYING BLOCKS AND GETTING A GOOD PRICE

Most building suppliers in your area will stock concrete blocks. The best price will be obtained either from a building supply company that stocks mostly masonry materials or from the block manufacturer if there is a plant near you. Small amounts of block are going to be more expensive because of hauling and handling charges. If you can, use a pickup truck and go directly to the manufacturer's

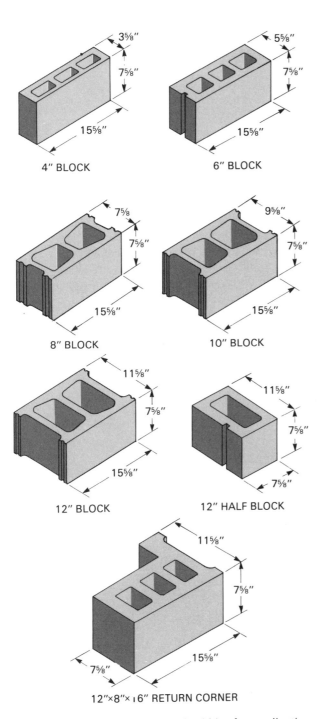

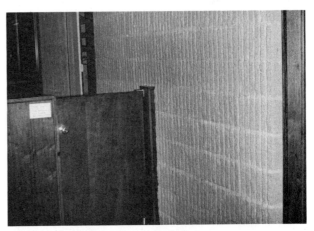

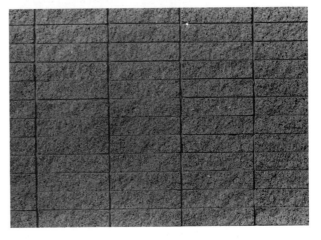

Blocks are available in several widths for applications from foundations to partition walls. There are even L-shaped blocks for foundation corners.

A wall of screen blocks provides good privacy but is much lighter and more open than a solid wall. Fluted blocks and so-called splitblocks are popular because of their strongly patterned, highly textured effect. There are many other special blocks to choose from.

yard. The block will be considerably cheaper as there will be no delivery charge.

There are also "seconds" for sale—blocks that may be slightly twisted, out of square, or imperfect in texture. Usually they are half the price of regular blocks. They are a good buy if you are building a porch or retaining wall, filling in a chimney, or doing some other work that does not require perfect blocks. You will have to ask for seconds, as the dealer or manufacturer rarely advertises them. If you don't know where the closest block plant is, check the yellow pages in the telephone book.

STARTING THE CORNER WITH DIFFERENT-SIZE BLOCKS

Most blocks are laid in either running or stack bond patterns, which are standard and are described in the brick chapter. Because blocks are large, they do not lend themselves to a lot of pattern variations. You can, however, achieve a variety of different appearances by raking out the joints or using a V joint.

The starter size of the block used on the corner depends on the width of the block being laid. Only 8" blocks can be laid half over the one beneath on the corner without cutting or using a specially made corner block. This is because 8" is half of 16", the standard length of a block. As each succeeding course is laid on the corner, it reverses and forms a half lap bond.

When a block is wider than 8", as in a 10" or 12" foundation wall, a specially made L-shaped corner block is used. The L-shaped corner block allows the block next to it to form the correct lap of 8". Different sizes of corner blocks are used for 10" and 12" walls, so be sure to specify to the supplier which size you want when ordering your blocks.

As previously mentioned, L-shaped blocks are made only for 10" or 12" blocks. All other width blocks, except 8", require a cut piece on the corner to start off the bond to achieve the 8" lap. The way you determine what length cut piece to use is to add the width of the block being laid to 8". For example, if you are going to build a 4"-wide block wall, the piece should be 12" long, which is the standard lap of 8" plus the 4" width of the block. If you are going to build a 6" block wall, then the cut piece should be 14". There are also 3"-wide blocks made, but it is very seldom that you will use them. Their starter piece would be 11" long.

PLANNING TO MAKE A PROJECT WORK IN WHOLE BLOCKS

Think ahead when designing a block project and try to make it work in whole blocks as much as possible without cutting. This not only looks neater, but saves a lot of time.

If you always think in multiples of 8", usually a project works without cutting. This is not always possible, but in most cases it can be done.

How L-shaped corner blocks are used. The corner block for the next course will run the other way, and the regular block will fit into it from the left.

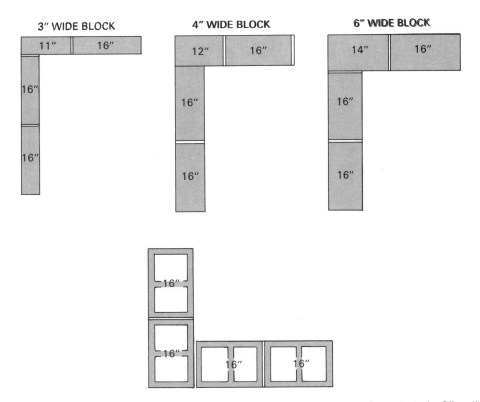

Proper length of cut corner blocks for walls of various widths: 8"+width of wall=length of cut block. An 8" wall, of course, requires no cut block, since 8"+8"=16", the length of an uncut block.

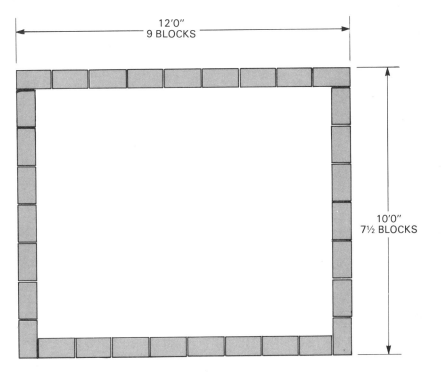

Layout of garden shed foundation with dimensions in multiples of 8", making cuts unnecessary at corners.

The same is true for the height of the project. The door, window, and wall height should all work in full blocks without cutting. In the illustration of a wall elevation, the door height is 7′4″, the window height is 4′8″, and the total wall height is 8′10″—all permitting full courses of blocks. The widths of the openings are also in multiples of 8″. You can see the advantage of planning your project to work in whole blocks.

Laying Out the Bond

There are three preferred methods of laying out the bond for a project. One is to use the mason's 6′ folding pocket rule and marking on the base every 16″ in length, which would be one block. Most rules have an identifying mark at these 16″ points, such as a diamond or red mark. Place your mark a little outside the actual wall line on the concrete footing or base so that it is not covered with mortar as you start to lay the blocks.

The second method is to use a 4′ level in place of the rule, as 48″ is three multiples of 16″, or three blocks. This is quick but a little risky if you don't match up each mark when you shift the level to mark again. However, it is a commonly used method on masonry jobs.

The third method is to use a steel measuring tape, and for a long wall it is the best method. Simply mark each 16″ to lay out the bond. A block wall that is 30′ long would take 22½ blocks exactly.

All of the methods will work. Select the one that fits the size of project you are building and is easiest for you.

ESTIMATING CONCRETE BLOCKS AND MORTAR

The simplest method of estimating your block needs is first to add all the wall lengths of the project together to find the total linear length. Since a standard block is 16″ long, there are three blocks to every 4′. This is a ¾-to-1 ratio. Multiply the total linear feet by .75, and the result will be the number of blocks on one course. Now you have a basis for figuring the rest of the project. Be sure to deduct for any openings, of course.

As stated before, each course of block including the mortar joint is 8″ high. Therefore, if you divide 8″ into the total height of the project or building (in inches), it will give you the number of courses high. Then multiply the number of blocks on one course by the number of courses high. This will give you the total number of blocks in the project or wall. If this is a foundation wall that is either 10″ or 12″ wide, don't forget that you will need one L-shaped corner block for every course on each corner. Deduct these from your total block amount so you don't have extras. The width of the block does not affect the quantity needed.

Estimating the Mortar

To estimate how many bags of masonry cement you'll need, allow 1 bag for every 30 blocks. Divide the number of blocks by 30 and this will give you the number of bags needed. If it works out to a fraction, simply round it off to the next higher whole number. The extra amount will not be that

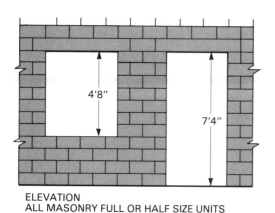

ELEVATION
ALL MASONRY FULL OR HALF SIZE UNITS

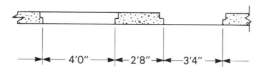

A wall planned so that all openings are in multiples of 8″ in both height and width, thus permitting the use of whole blocks throughout.

The top of a block has wider edges than the bottom, providing broader surfaces for applying the mortar.

much and will allow a reasonable amount for waste.

If you are using 40-pound bags of packaged dry-mix mortar, it takes about 1 bag for every 17 blocks. Remember the sand is already in this mix, and all you do is add water.

If you are going to use portland cement and lime mortar, (based on 1 part cement to 1 part lime to 6 parts sand), 1 bag portland cement to 1 bag hydrated lime with 42 shovels sand will lay about 62 blocks. Any of these mortar mixes are all right. Select the one you want to use.

A ton of sand is needed for every 8 bags of masonry cement. Allow a little extra if you're piling it on the ground.

Since the number of bags of masonry cement has already been determined, simply divide that figure by 8 to determine how many tons of sand you need.

The amount of water needed depends on how dry or wet the sand is and how stiff you want your mortar to be. This is a judgment factor at the time you mix the mortar. I recommend, however, that you always mix your mortar a little stiffer than for bricks, because the blocks are heavy and you lose some of the mortar down the holes or cells of the blocks.

When you estimate your project, be careful not to overfigure, as it is not only a lot of work to return materials, but the building supplier generally will charge a refund penalty, usually at least 10%.

When picking up your masonry cement, it is best to buy only a portion if the amount is going to be considerable, and then get the rest as the job progresses. This will assure you of having good fresh cement without the worry of having it get hard in the bag over a period of time.

HOW TO LAY CONCRETE BLOCKS

Laying concrete blocks is somewhat like laying bricks, with some exceptions. Blocks are larger, heavier, and more delicate than bricks. They are also more difficult to cut than bricks, because of the hollow cells.

Concrete blocks should be kept dry as much as possible before laying. If they are heavy with moisture they will sink into the mortar bed and you will not be able to keep them level, plumb, or to the correct height. When a highly absorbent brick is dry, it is advisable to dampen with water; this is never true with concrete blocks. The drier the better. Moisture can cause blocks to expand slightly. If they are laid in the wall while wet, they will shrink as they dry, and this could cause some minor cracks in the mortar joints. Cover them until you are ready to use them with some type of plastic or tarpaulin.

The same tools recommended for brickwork will do for blockwork. Refer to the tool list in the brickwork chapter.

Determining the Top from the Bottom of the Block

A concrete block has a top and a bottom. The top of a block has wider edges, including the interior webs, which make it easy to detect. The bottom of the block has larger cell openings or holes and thinner edges. The reason for laying the block with the widest edges up is that it is easier to apply the mortar and there is less chance of losing mortar in the open cells. This may seem a minor point, but you will find if the block is not laid correctly, it is more difficult to build a block wall.

Laying the First Course

Lay out the bond to work in whole blocks before spreading any mortar. Remember to allow ⅜″ between blocks for the mortar joints.

The mortar bed joint under the first course of block must be full to prevent water from leaking through. When leveling a block by tapping with the trowel blade, tap in the center to avoid chipping the face.

Since a block is made square in both directions, you can level across the first block; when it is level, it will also be plumb. This is the only time that a block is leveled across the top for plumb. From this point on, the wall face is plumbed vertically with the level.

Once you have two corner blocks in place, attach a line to the top outside edge of the wall with line blocks or a line pin and a nail, as when laying bricks.

It takes practice to apply mortar to blocks. The simplest method is shown in the photographs.

To prevent the mortar head joints from falling off when you lift the block to lay it in the wall—a problem many beginners have—press down with the trowel from the edges to the center of the block with a pinching motion.

Only the outside edges of the end of a block need to be buttered with mortar. This provides an open center area where you can pick up the block without getting your fingers constantly in the mortar, which will in time make them sore.

When you have finished laying the first block course, cut away the mortar from the bottom next to the concrete and smooth the joint flat with the trowel blade. This bottom bed joint should never be tooled, but always smoothed with the trowel regardless of the joint finish.

I recommend wearing a pair of work gloves when laying block.

Laying out the first course of blocks. The 4' level is being used as a gauge; three full blocks with ⅜" spaces for mortar run exactly 4'.

The mortar bed under the first course should be full to keep water from leaking through. Smooth it approximately level, without furrowing.

Level the block, but don't settle it too far into the mortar until the height has been checked. You can also use the level across the width of the block to check for plumb when laying the first course, but subsequent courses should be plumbed vertically.

Check the course height with a folding rule. With the mortar bed, it should be 8".

Lay a block at the other end of the wall and make sure it is level with the first. Here the point of the trowel is supporting one end of the level because the distance is slightly greater than 4'. For still greater distances, put a straight piece of lumber across the blocks and set the level on it.

Attach a line to the top outside edge of the wall, using line blocks as explained in the brick chapter. Adjust the ends of the blocks to line up.

The simplest method of applying mortar to head joints is to pick up about half a trowelful of mortar, set it on the trowel with a downward snap of the wrist, and then roll it onto the end of the block with a pinching motion toward the center to make it stick when you pick up the block. Only the outside edges need to be mortared, leaving the center free for your fingers.

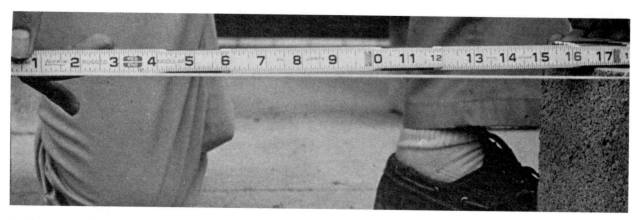

For the closure block—that is, the last one in the course—allow about 16½", to permit a mortar head joint on each end of the block.

Trim the mortar at the bottom of the first course and smooth the joint flat with the trowel. Never tool this joint, even if the other joints in the wall are to be tooled.

When laying the closure block, apply mortar to ends of the blocks on either side as well as to both ends of the closure block itself. Lay it in position and line it up carefully. Point up any holes in joints, trying not to smear edges.

Laying the Rest of the Wall

Applying mortar is done a little differently for blockwork than it is for brickwork. As a rule, mortar is not usually spread across the center of the webs, as it is not needed. The best method of applying the mortar is to pick up a modest trowelful of mortar, set it on the trowel with a slight downward snap of the wrist, and swipe it on the outside edge of the block. Don't hold the trowel perfectly vertical or the mortar will fall off to the outside. It will also fall off if you hold the trowel at too great an angle over to the inside. The correct angle is just a little lean to the inside of the block as shown in the photograph. Reverse this for the opposite side. This will take some practice before you get the hang of it.

Laying Blocks to the Line

The line should be even with the top edges of the blocks at all times if a straight wall is to be built. Pick up the block at both ends for balance. Don't lay the trowel down each time you lay a block and then pick it up again to adjust the block to the line.

For the next course, apply mortar along the outside and inside edges of the blocks already in place. You'll soon learn how to do this without dropping the mortar on the ground or into the cores.

Check the height of each course with a rule. It should be in multiples of 8".

Plumb the ends of the wall, called the jambs. Hold the level snugly against the bottom and make the end block fit flat against it.

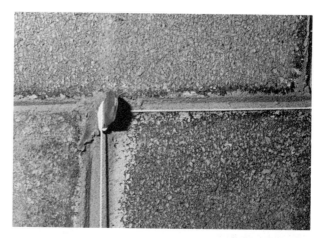

Attach line to the end or corner blocks by either the line-pin or the line-block method, as when laying brick.

Once the end blocks are leveled and plumb, use the level as a straightedge along the front of the wall to correct any minor misalignments.

Pick up the block at both ends. Don't put the trowel down; develop the habit of grasping around it to pick up the block. This saves a lot of moves in laying blocks.

Instead, hold the trowel in your hand and grasp around it to pick up the block. This tip will save you a lot of time and finishing.

Lay the blocks in the mortar bed, position them, and trim off any protruding mortar as shown in the photographs. When cutting off protruding mortar,

hold the trowel at a slight angle to prevent smearing the face of the block.

After the mortar joints have stiffened enough that they will not smear, they should be tooled. You can determine if they are stiff enough by pressing your finger or thumb into the joint. If it leaves an

Cut off excess mortar, holding the trowel at a slight angle to prevent smearing the block or dropping the mortar on the ground.

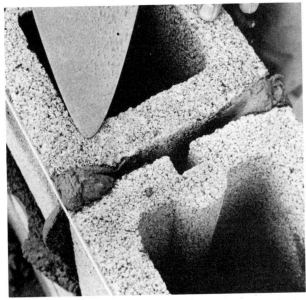

Lay the block in gently to keep it from sinking too far into the mortar. The mortar should squeeze out to form well-filled joints as shown.

Don't lay the blocks actually to the line, but about 1/16" back from it, so that no block will touch the line and push it off dead straight.

impression, the mortar is ready. Tool the head joints first, filling up any holes as you go. Picking up mortar with the striking tool was described in the bricklaying chapter.

Before brushing, cut off any protruding mortar that remains from the striking process. Do this by holding the trowel on an angle and shaving the excess off. Be careful so as not to smear the mortar on the wall.

Let the mortar dry enough not to smear, then brush the wall lightly with a medium bristle brush. Never use a house broom; it will ruin the struck finish. If necessary, rebrush and restrike a second time to make a real neat job.

Although there are variations of blockwork techniques, the ones so far described are for basic blockwork and should serve your needs.

Cutting Blocks

You will need to cut blocks for some projects now and then. In the bricklaying chapter, cutting with a hammer was described. Blocks can be cut with a hammer in about the same manner. However, they can be cut more accurately with a brick set chisel. Because blocks are more fragile and have large open cells, they are more difficult to cut than bricks.

An effective method of cutting is shown in the photographs.

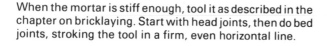
When the mortar is stiff enough, tool it as described in the chapter on bricklaying. Start with head joints, then do bed joints, stroking the tool in a firm, even horizontal line.

Before brushing the joints cut off any excess mortar left by the striking tool, being careful not to smear. When the mortar is fairly dry, brush it with a medium-bristle brush.

A small trowel is very useful when neatening up an inside corner. Be sure to rebrush after using the trowel.

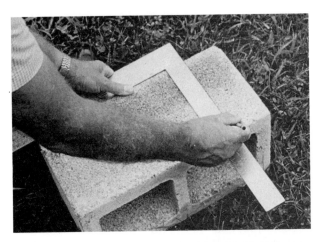

Cutting a block. First mark the cut with a pencil and square. Then score along the marks on each side of the block with the brick set chisel, hitting hard enough to make an impression but not to break through. Then set the block on its bed side and strike along the scored line from the top. Trim the cut end with care, using light vertical pecks with the hammer.

33
CONCRETE WORK

Concrete work is by far the most durable, versatile, and economical of all masonry materials. Materials like block, brick, and stone are necessary for a good footing or base and as such are the heart of any masonry job.

In addition, there is information on a variety of uses of concrete (such as patio blocks, exposed aggregate finish, and creative projects) along with how to color concrete and problems that can occur and how to correct or prevent them. As you read through this section and prepare for your concrete project, keep one important thing in mind—once you have poured and finished the concrete, the only way it can be changed is to tear it up completely and start over. When laying brick, block, or stone, the mortar can be removed and rebuilt, but with concrete you will have to get a jack hammer and the correction process would not only be expensive but backbreaking.

Concrete is one of the most interesting and flexible of all building materials because it can be molded or formed into almost any shape. Adding to

that, its super strength and long life make it one of the most popular of all building materials.

Materials used to make concrete are portland cement, some type of fine aggregate, such as sand, and a coarse aggregate such as gravel or crushed stone mixed with water. When water is added to the mix, it causes a cement paste to form. This paste fills all of the voids between the cement, sand, and stone aggregate, making a solid mass when it has cured.

Aggregates should be reasonably clean and free from earth or any type of vegetation, loam, or silt. If you buy your materials from the building supplier, as a rule they are clean because the sand and stone are generally washed with a water spray before they were delivered from the major supplier. Where you run into problems is trying to save money and buying ungraded or unwashed sand from a pit or bank. It is also worth mentioning that seashore sand has a high concentrate of salt in it and never should be used to mix concrete. Sand for concrete is more coarse than sand used for brick or block mortar and particles will vary from dust to those approximately ¼" in diameter. Check with the building supplier and he will direct you to the proper sand for your concrete. The finer the sand, the more portland cement will be needed for a mix.

Generally, water used to mix concrete should be suitable for drinking. Silt, mud, or organic materials in the water will seriously affect the ultimate strength of the concrete, so be selective and don't use muddy or old standing water.

MIXING AND STRENGTHENING CONCRETE

Although there are a variety of types of portland cement available, stay with Type 1, which is normal or average. Portland cement is packaged in 94-pound bags that hold 1 cubic foot of dry material. The bag has a plastic lining to keep the cement dry and at full strength since cement is quickly damaged by moisture and will prematurely set. Always keep the bag closed and store off the floor or ground.

An important rule or law to remember when mixing concrete is called the "water-cement ratio." As previously stated, the cement paste unites with the aggregates to form the hardened mass we know as concrete. Therefore, the quality of the cement paste is a very important factor in producing good concrete. If you add a higher amount of water to the mix than is needed, while it is true it will mix easier, it will also weaken the concrete. The basic rule to follow is to use as small an amount of water in proportion to the cement as possible and still allow the ingredients to blend together. The common practice of adding a lot of water to concrete to make it run in a form simplifies placing it, but very seriously weakens it. If your concrete is mixed correctly, there should be some effort to move it in and around the forms. A nice gray color should be present and it should have a creamy texture with a topping of cement paste after floating with a straightedge.

There are two commonly used formulas for concrete that should serve all of your needs. For footings, use a 1:2:4 mix. This means 1 part portland cement to 2 parts sand to 4 parts crushed stone or gravel with the proper amount of water. For concrete slabs, floors, etc., or where you want a good smooth finish, use a 1:2:3 mix. This consists of 1 part portland cement to 2 parts sand to 3 parts crushed stone or gravel with water. These two formulas are given if you are going to mix your own concrete.

If you are ordering ready-mix concrete from the plant, there are two different methods of ordering you may use. One is to specify the number of bags of portland cement in one cubic yard. For example, a footing mix does not have to be quite as rich as a slab or floor mix because larger stones are used; therefore, a normal footing mix contains 5 bags portland cement per cubic yard and is a little cheaper than a rich slab mix. A slab or floor mix contains 6 bags portland cement per cubic yard. The other method used many places in the country is called a "prescription" mix. This is where you specify the psi (pounds per square inch) strength of the concrete. If your concrete dispatcher uses this method, keep in mind that a 2,500 psi mix will be used for a footing and a 3,000 psi mix will be used for finish concrete. Both methods allude to the same end results.

The addition of steel reinforcing rods (commonly called "rebar") and concrete reinforcing wire will greatly increase the strength of concrete. Both types of reinforcement are available from your building supplier. Rebar is used when you are pouring concrete that spans an overhang or where greater strength is required. Reinforcement wire is used to strengthen flat work. They can also be used together for maximum results.

Concrete will harden overnight enough that you can walk on it, but it takes about 28 days to reach testing strength. It is better to mix small amounts (1 yard or under) with a utility mixer at the jobsite, but on larger amounts, I recommend ordering ready-mixed from the plant.

One of the most important advances in concrete development in recent years is air entrainment, which improves its workability and durability and results in a high resistance to damage occurring from severe frost or freezing actions. It is especially recommended by concrete manufacturers for flat work such as slabs, walks, floors, etc. Properly proportioned air-entrained concrete contains less water per cubic yard than regular concrete and results in a more weather resistant, blemish-free surface. It also will not crack as easily as regular concrete. The reason it works so well is that it has millions of small air bubbles entrained throughout the concrete. These air bubbles act as cushions and flex with stress without cracking. It sets or hardens a little slower than regular concrete and is more foamy when poured. It also costs slightly more than regular concrete. I would consider using it if pouring concrete in a cold climate or where a lot of ice or snow may accumulate. Air-entrained concrete holds up very well against the salts that are used on roadways and pavements during the winter months. It should be ordered ready-mixed from your concrete supplier.

Be wary of using any admixtures in your concrete that would accelerate or retard the setting time, as they can affect the strength. The ideal temperature for curing concrete is about 72°F. Try to schedule your concrete in nonfreezing weather.

HOW TO ESTIMATE AND ORDER CONCRETE

The first step in estimating concrete is to calculate the total area in square feet by multiplying the length by the width of the area. Next, multiply the square footage by the thickness in feet. The result will be the total volume of concrete in cubic feet. Convert the cubic feet into cubic yards by dividing by 27, as this is the number of cubic feet in a cubic yard. This is fairly simple arithmetic.

To estimate the amount of concrete for a footing, simply find the total number of lineal feet around the walls, including any porches, retaining walls, etc., that are a part of the foundation. Then multiply the length of the footings in feet by the thickness in feet by the width in feet. Now divide this number of total cubic feet by 27 to find the number of cubic yards needed. Same process and same results. Always recheck your math, as it is easy to make a mistake. As a rule allow about 5% more than you really need to fill in any places that may have sunk a little.

The formula for estimating concrete is as follows:

$$\frac{\begin{array}{ccc}\text{length} & \text{thickness} & \text{width} \\ \text{in feet} \times & \text{in feet} & \times \text{in feet}\end{array}}{27} = \text{cubic yards}$$

You may find it easier to express inches as fractions and decimals of a foot in your calculations. The following table illustrates these:

Inches	Fractional Part of Foot	Decimal Part of Foot
4″	1/3	0.33
5″	5/12	0.42
6″	1/2	0.50
7″	7/12	0.58
8″	2/3	0.67
10″	5/6	0.83
12″	1	1.00

If you are going to buy transit mix, most concrete suppliers will use a slide rule to determine how much you need, basing their calculations on the critical measurements you give them.

You have several choices to make when buying concrete or concrete materials. You can mix your own, as stated, if the amount is less than a cubic yard. If the amount is very small, such as around 100 or 200 pounds, a dry ready-mix is available to which you only have to add water. Although it is an expensive way to buy concrete, it is economical enough for small amounts. An 80-pound bag of this type of mix will work fine for that fencepost or mailbox you want to make secure in the ground.

Concrete mix to which you only add water is convenient for very small jobs but too expensive for larger ones.

Have an extra place ready to pour when mixing or buying concrete, as it usually is impossible to mix exactly enough with none left over. You can usually think up a good use for the excess—a short length of sidewalk, a concrete base for the garbage cans to sit on, or that spot that always gets muddy at the end of the driveway.

Another tip is to order your concrete a couple of days ahead and have plenty of help lined up if it is a sizable amount. Avoid by all means ordering concrete after regular working hours or on Saturdays, as there is surely going to be an extra charge. As a rule the driver is allowed a set length of time to unload the concrete on the job, and you will be charged for any time longer than that. Inquire when you order how much time is allowed on the job.

Another point to consider is the amount you need. It costs the supplier as much to send a truck with 1 cubic yard of concrete to the job as to send one with 8 cubic yards. Mileage from the concrete plant is also a big factor in how much the concrete will cost. You can't get around all of the variables that affect expense but planning ahead and making the best use you can of the truck is worth considering. This is why in most cases it is a lot cheaper to mix anything under a cubic yard yourself.

One last tip on ordering concrete: Check the weather report, and if it calls for rain, you may want to postpone delivery for a better day. When you're pouring footings or concrete that will be under the ground, you don't have to be too fussy about troweling; if there is a light shower it won't hurt the concrete. However, if you are putting in a concrete slab or deck that is exposed to the elements, the weather must be taken into consideration if you want to do a good job.

TOOLS AND EQUIPMENT YOU WILL NEED

A number of the things you'll need are probably hanging around your house right now. The few you will have to buy or borrow are not expensive and are carried by almost all building suppliers.

- Bricklayer's trowel (It can be an old one.)
- Mortar hoe
- Hose
- Square shovel (Round-point can be used but is not as handy to pick up stones with.)
- Steel tape or rule
- Bucket (5-gallon)
- Ball of line
- 2' square
- Edger
- Steel finishing trowel
- Wood and metal floats
- Screed or strike-off board
- Jointer
- Bull float (You can make a wood one or rent a metal one.)
- Level
- Rake
- Brush or soft-bristle broom

The steel finishing trowel, edger, jointer, and bull float are probably the tools you will need to buy, borrow, or rent.

POURING A FOOTING

There are two types of footings you can pour or place, the trench footing and the formed footing.

In a trench footing, the sides of the trench serve as forms for the concrete. This is the simpler method to use. Try to dig the trench walls as vertically straight as possible and keep the bottom reasonably level. Before pouring, drive short steel rods or sharpened wood stakes about 5' apart in the center of the footing to serve as leveling points for the concrete. As the concrete is poured in the trench, any wood stakes should be bumped loose and removed so that they do not rot and leave hollows, but the steel rods can be left in place. Remember the accepted rule for footing size: twice as wide as the block to be laid on it and as deep as the height of one course of block, which would be 8" in most cases.

A formed footing is required when rock or very hard ground is encountered. First, drive wood or steel stakes at the corners of the structure or project in line with where you want the footings to be placed. Make sure the stakes are level by using a long straight board with a mason's level laid on top or by using a garden hose as a level. Mark a level line on the stakes and stretch a line tight. The main thing is to build the footing forms strong enough not to bow or give once the concrete is placed in them. Set up the outside form first. Make the tops level with the nylon line attached to the corner stakes. Use ¾" plywood or 2"-thick boards such as 2×8s or 2×10s. Hold them in place with sharpened 2×4 stakes driven and nailed on the outside edge of the forms. Make sure that the tops of the stakes are level or a little lower than the top edge of the form so that they do not interfere later with leveling the concrete off with a screed. Space these 2×4 stakes about 3' apart for good support. Tack some 1×2 wood strips across the tops of the forms about every 6' to prevent them from spreading. If the sides of the concrete will be exposed after the forms are removed, it is a good idea to coat the forms with a light coat of clean oil, preferably light motor oil or form oil.

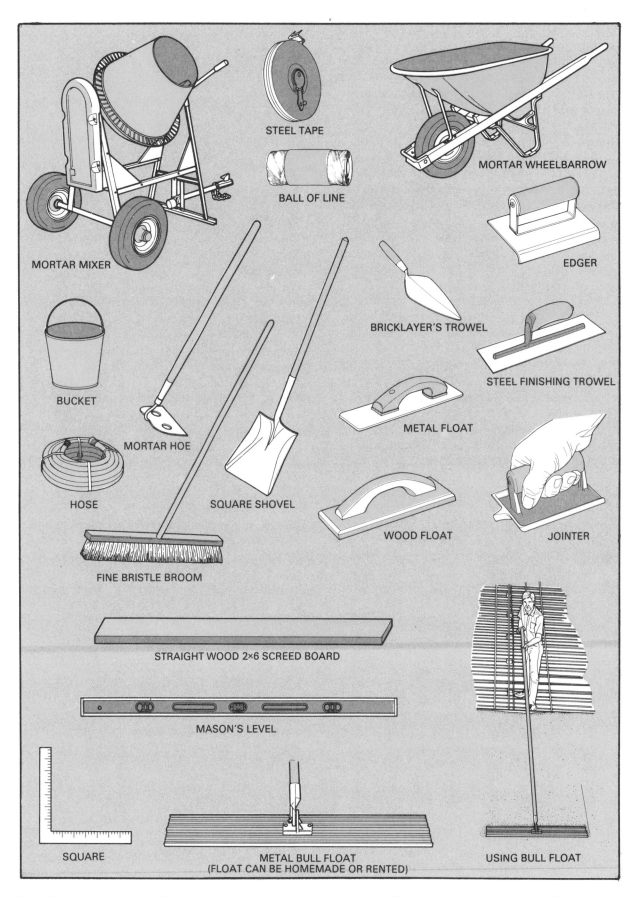

STEEL TAPE

BALL OF LINE

MORTAR MIXER

MORTAR WHEELBARROW

EDGER

BRICKLAYER'S TROWEL

STEEL FINISHING TROWEL

BUCKET

MORTAR HOE

METAL FLOAT

HOSE

SQUARE SHOVEL

WOOD FLOAT

JOINTER

FINE BRISTLE BROOM

STRAIGHT WOOD 2×6 SCREED BOARD

MASON'S LEVEL

USING BULL FLOAT

SQUARE

METAL BULL FLOAT
(FLOAT CAN BE HOMEMADE OR RENTED)

Many of the tools you'll need for concrete work you probably already own. The others are not expensive and can usually be rented.

Mix concrete for a footing in the proportions of 1 part portland cement to 2 parts sand to 4 parts gravel or stone. I suggest you dry-mix the materials first in the wheelbarrow or mortar box, then pull all of the mix to one end and add a couple gallons of water. Work the dry materials into the water little by little with the hoe, rather than trying to mix all of it at one time. This is much easier on the back muscles. Continue to add water as needed until the concrete is mixed to the consistency you want. As a rule this should be so that the concrete will slide off the shovel but not run off. You should be able to smooth the surface out easily so the aggregates are not showing. Every particle of the mix should be coated with cement paste. The concrete should be easy to handle and place in the forms, but not so liquid that it runs. This is the secret to a good strong mix. Many people cut the strength by adding too much water to make it easier to handle. A little practice in mixing and you will soon get the hang of it.

If you are going to use a utility mixer, one of the problems is that the materials stick to the sides. Try this for better results. First put about 2 gallons of water in the mixer, then half of the crushed stone. Have the mixer turning while adding these materials. Next put in about half of the portland cement and sand. Add a little water and then the rest of the materials to make a full batch. Be careful about standing and looking directly into the mixing drum as it is turning, as the concrete sometimes slops out and invariably it is going to be in your face or on your clothes. For most utility mixers, a good-size batch would be about 3 shovels portland cement, 6 shovels sand, and 12 shovels crushed stone.

Pour the concrete into the trench or formed footing and tamp it level with a garden rake. Never trowel smooth the top of a footing, as if you do the mortar joint under the first course of block will not bond to it as well. The better the footing is for level, the easier it is to lay that all-important first course of block on it. If wood stakes were used as level

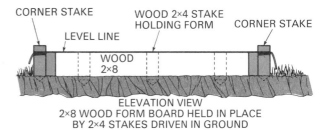

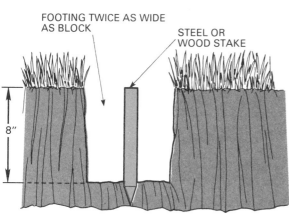

TRENCH FOOTING

Concrete can be poured directly into a trench footing. Stakes are used to provide leveling points. The sides of the trench should be as vertical as possible.

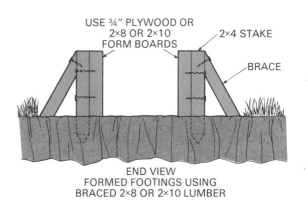

A formed footing is made of plywood or lumber with 2×4 stakes and braces. A better finish will be obtained if the forms are coated with oil before pouring the concrete.

points in the center of the footing, they should be removed as the concrete is poured, as otherwise they will rot out later.

FORMING FOR
A SIDEWALK OR SLAB

A frequent job is making a form for a sidewalk, slab, or pad for something to sit on. The standard concrete slab for walks, slab, etc., is about 4" thick. As a rule, wood 2×4s work fine, but their actual width is 3½", not 4". Excavate a little deeper than the actual thickness of the 2×4 to overcome this, or use rough sawmill lumber, which is 4". Install the forms the same as for formed footings by driving level stakes into the ground at the ends of the slab and attaching a line to the stakes. Then the 2×4s are set into position to the line and nailed to 2×4 stakes to hold them firmly. There is a special double-headed nail that works great for forming; it makes the job of removing the nails later a lot easier. Square the corners with a carpenter's framing square and brace them. If the project is a sidewalk or patio slab, it is a good idea to place temporary dividers at intervals of about 10' and put an expansion joint there to allow for movement.

If the soil beneath which you are going to make the pour is firm and compact, you can spread a thin layer of crushed stones or simply lay some plastic on the ground followed by concrete reinforcement wire for strength. I recommend plastic, as it helps the concrete to dry more evenly and the water is not absorbed by the dry ground. Be sure to dig out any grass, weeds, or high spots before stretching the plastic on the ground. Fill in any low spots with tamped soil or crushed stone.

Allow a little pitch or slope, about ⅛" of width per foot, so water will run off. Make one edge a little higher than the other and that will solve this problem. Always slope the walk or slab away from the house so you don't have water standing. The best mix to use for slabs, walks, or finished concrete that has to be troweled smooth is the 1:2:3 mix described earlier.

Concrete forms can also be built on curves by using plywood and bending it around the curve. Because of the curve, more stakes will be required to hold the plywood in place. You may also have to use thinner plywood to make the curve.

POURING, FINISHING,
AND CURING CONCRETE

Now that you are ready to pour the concrete, have some help lined up as previously stated, and a good straight wood 2×4 or 2×6 to use as a screed or strike-off board.

Trucks that deliver concrete have a chute on the rear that can be swung into position, so that the concrete can be poured directly into the form. For most small jobs, however, the simplest method is dumping the concrete into the form from a wheelbarrow. Starting at one corner or end, being careful not to rest the wheelbarrow against the forms, dump the concrete into the formed area and at the same time work it ahead and against all sides of the forms to fill out all areas with the shovel or rake. Lift the reinforcement wire off the plastic with the rake as you pour so that it will be firmly imbedded in the mix.

Don't dump the concrete in separate piles, but always right against the previously poured mix,

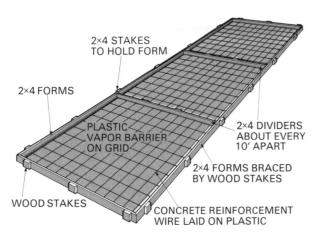

A form for a sidewalk or slab. Since a milled 2×4 is actually only 3½" wide, you will have to excavate a little below the form to achieve the usual 4" thickness.

Forms for concrete can be built to make curves. Plywood, in thicknesses less than ¾" if necessary, works well for the sides of the form.

working it to the ends. Tapping on the edges of the form with a hammer will help to settle the concrete fully. The same effect can be achieved by slicing along the inside edge with the blade of the bricklayer's trowel. Fill in any of the concrete that settles at this point.

As soon as one block or area is poured to allow for working room, so as not to interfere with the pouring operation, strike off or screed the top of the slab. This is done by moving the wooden screed in a sawing motion back and forth, letting it rest on top of the edges of the forms. A third person can fill in any low spots or pull with the rake any buildup of concrete, making it easier to screed. The screeding

will force all of the aggregates down into the mix and bring the cement paste up on top. This is called the "cream" and has to be on top in order to trowel the concrete smooth later.

As soon as some of the surface water evaporates, float the concrete surface with a wood or metal float to work up more of the concrete paste and make sure that there are no aggregates exposed. On large areas, a bull float that has extension handles can be used. For walks, porches, and small slabs, however, you will not need one.

The edging should be done next. Run the edger around all of the edges with a firm steady motion to make the edge. If you encounter a place that is not

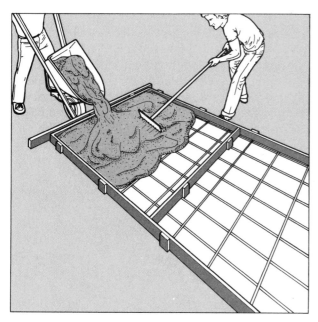

Filling a form from a wheelbarrow. Start at a corner or end, and work the concrete into the corners and against the sides of the form as you go.

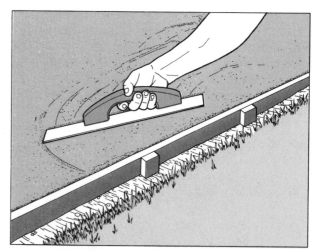

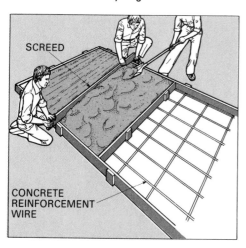

Screeding. The wooden screed is moved in a sawing motion along the tops of the forms, leveling the concrete and forcing the large aggregates down below the surface.

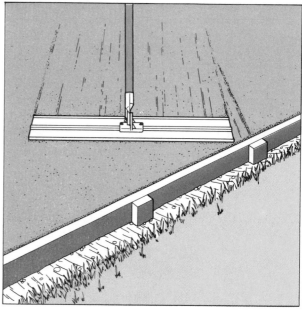

Small areas can be floated by using a hand float, but a bull float that has an extension handle will be a big help for larger areas.

filled up perfectly, add a little of the concrete paste from the top of the slab and re-edge it. It is very important that the edging be well done and all voids be filled as soon as possible after floating to get a good edge. Be careful not to let the edger dig in too much; hold the front of the tool up slightly and let it glide on the edge.

If there are dividers or expansion joints between slabs, run the jointer (groover) across against a guide strip or board. This is the familiar groove one sees in sidewalks or large sections of concrete paving.

The next part of finishing, an all-important one, is the first troweling with the steel trowel. The trick is to know when to trowel the first time. Whatever you do, don't take a break now and grab a sandwich and coffee in the house. Stay with it. Your wife can bring you a snack. The troweling cannot be done until the water sheen leaves the surface and the concrete sets a little. On a cool day you will have to wait a little longer; on a hot windy day the time will be shorter. Keep checking, and as soon as you can trowel the surface without sinking in or leaving a lot of trowel marks, it is ready. The first troweling is

very important, as you want to smooth the surface as much as possible and leave a creamy smooth texture over the entire surface.

Let the concrete set until it stiffens again and the paste becomes stiffer. You will have to experiment a little to get the feel of this, as experience is the best teacher. Retrowel again, trying to get all of the trowel marks out this time and leaving a smooth slick surface. Make your troweling motions with the trowel slightly raised on the front to prevent digging in and with pressure exerted downward to glide over the surface. If you want a still smoother surface, let it set and retrowel one last time. It is easy to tell when the concrete is too stiff to trowel, as sand will start to come up to the surface.

As you can see, the timing in troweling for a good smooth surface is very important and no words can teach you exactly how this is done—only trial and error on your part. It is not hard to do, but requires constant attention to the job. If you have to reach out on the slab, lay a board or a flat piece of clean plywood about 16" square and rest your knee on that. Be sure to retrowel that area when you move the board.

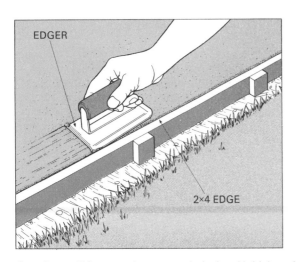

An edger will leave a clean, rounded edge. Hold the edger so that it doesn't dig in as you draw it along.

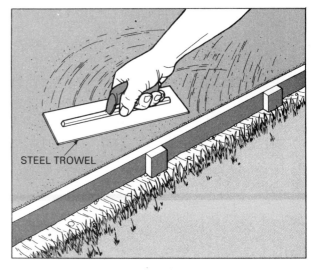

The steel trowel has to be used at just the right time—when the water has evaporated from the surface and the concrete has set slightly but can still be smoothed easily. As the concrete continues to set, it is troweled again, and perhaps even a third time for a really smooth surface.

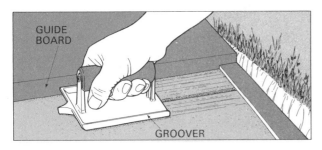

The groover or jointer is used to make expansion joints. Use a board as a guide to keep the joint straight.

Many people like a slightly rough surface on sidewalks or exterior slabs. This can be achieved by waiting until after the first troweling and then lightly brushing a broom over the surface. The texture will depend on the stiffness of the broom used. I recommend a medium-bristle floor broom, with a long handle to minimize making pockets or rough spots. You can make the pattern wavy or straight, but only gritty and not so rough that the concrete is crumbly. Complete the job by dressing up along the edges. If the surface is too dry, dampen the broom a little with the hose.

CURING

If concrete is to attain its maximum strength, proper curing is important. The weather is going to be an important factor in the curing. If the weather is cool and damp, then it will take longer. Don't be fooled by appearance. The concrete, if it is a finished slab or walk, should not be walked on for at least three days, especially on the edges.

The idea is to keep the concrete slightly dampened so that hairline cracks do not appear. If you are around the house, then dampen with a fine spray from the hose a couple of times a day and this will do the job. If you are going to be away and the weather is warm, wet burlap or a sheet of plastic wetted down works well. Straw also works well if kept wet. A good rule of thumb to follow for curing time is about one week to be on the safe side. On construction jobs, concrete has to be 28 days old before strength testing is done.

Wood forms should be removed as soon as possible after the concrete has set enough that it will not move or sag. This allows the edges to cure along with the surface. I generally remove them the day after placement. Any fine holes or voids on the edges can be filled up by mixing up a little portland cement and water into a mortar consistency. Scrub the forms off as soon as possible after removal with water and a stiff brush so that they can be used again.

The following series of photographs show how a typical concrete sidewalk is placed.

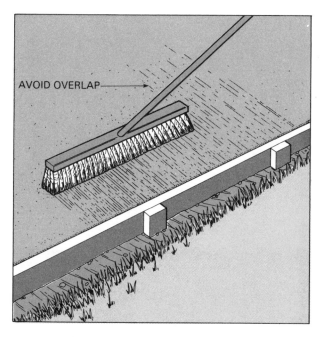

After the first troweling, the concrete can be brushed to make a slightly rough nonskid surface. Brush in long, even strokes without stopping in midstroke; don't overlap your strokes, just butt them.

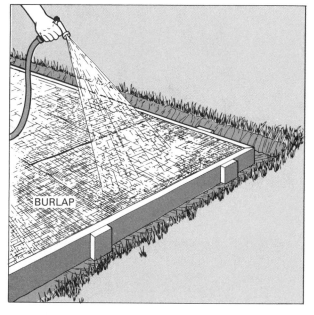

Concrete should be kept damp while curing. If you won't be around to spray it a couple of times a day, cover it with wet burlap or plastic sheet to hold in the moisture.

Wood 2×4 forms braced into position with concrete reinforcement wire, ready to receive the concrete.

Concrete is screeded even with edges of the form with a 2×4 screed board.

Ready-mix concrete arrives on the job and is ready to be poured into wheelbarrow through chute on rear of truck.

Longer widths of the walk are screeded with a long screed board. This sidewalk is next to a backyard swimming pool.

Concrete is placed into forms from wheelbarrow and worked in position with the shovel and rake.

Divider strips were placed about every 10' with expansion joints to allow for any movement. When this point is reached with the placement of the concrete, the form board is removed and filled in with concrete against the expansion strip.

A wood or metal float is used to bring the cement paste on top after the screeding process. This should not be done immediately, but as soon as the free water on top evaporates enough that the cement can be smoothed out.

Any place where the concrete reinforcement wire works its way to the surface will have to be cut off with cutting pliers and filled in with concrete.

The edges of the walk are rounded off with an edger tool as soon as the floating is done to fill all voids. The same is true for the concrete next to the expansion joint that runs across the walks.

The first troweling is done with the steel trowel after it has set a little but still has a soft cement paste on top.

A bull float or flat trowel with a long handle is used to reach across the walk and smooth the surface.

Final troweling is done after the concrete has set enough that the trowel marks are not going to be very noticeable. It is almost impossible not to have any trowel marks in the final job.

FORMING AND POURING CONCRETE STEPS AND PLATFORM

A long-lasting set of steps and porch can be formed and poured of concrete to suit any size needed. Build the foundation for the porch and steps of concrete blocks below freeze depth and approximately level with existing grade line. This is a good place to use old blocks or seconds from the block company that are cheaper than new or first-grade blocks.

Although ¾" plywood can be used for forms, I prefer to use 2×8 boards, as they are heavier and will not bow as easily. In addition they are 7½" high, which is the height of each of the finished steps.

Form the porch and steps as shown in the illustration. The platform (top) can be as wide or long as you want it and not necessarily the same width as the steps. However, I have the steps and platform the same width in the drawing to simplify things.

I used 2×4 braces sharpened on the ends and driven in the ground securely to hold the forms in position. This is a must to brace all forms so there is no possible movement when the concrete is poured. Double-headed form nails, which are available from building suppliers, should be used to nail the forms in place. They are easily removed when it is time to take the forms off.

I prefer a tread 12" deep and a riser 7½" high. A brace will be needed from the bottom step to the top edge of the platform. Nail this in about the center of the steps. It prevents the steps from bowing out of position as the concrete is poured. Use a 1×3 with a double-headed nail.

Oil the inside of the forms to assure a smooth finish when the forms are taken off.

I use a 6-bag mix to obtain a smoother finish. You can reduce the amount of concrete needed by filling in the porch area with stones, old broken bricks, blocks, etc. Just be sure to keep all of the fill materials at least 12" away from any edge of the form to make sure that the concrete is filled out against the forms.

As the concrete is placed in the form, float the steps with the wood or metal float tool until it is level with the top of the form.

After the concrete has dried enough and free water has evaporated from the surface, edge the steps and platform with the edger tool to form a slightly rounded edge. Complete the finishing job by steel troweling until smooth or if a coarse finish is desired, brush with a medium bristle brush.

The forms should be removed carefully before the concrete has completely hardened and the edges filled and smoothed with the trowel. Usually if the weather is nice and above freezing, the forms can be removed in the late afternoon and still finished along the edges. This, of course, depends on drying conditions.

Remove all nails from the forms and scrub clean with water so they can be used another day. Dampen the surface of the finished concrete with a fine spray of water about every 8 hours to promote curing. A sheet of plastic laid on top of the concrete overnight will retain the moisture; however, remove it in the morning to allow drying.

OTHER CREATIVE USES OF CONCRETE

Exposed Aggregate Finish

Exposed aggregate concrete is very popular for a rustic effect where the homeowner wants something a little different. It blends in with the natural surroundings and is a good choice for patios,

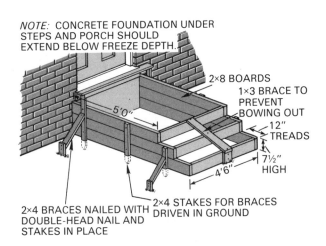

Proper construction of forms for concrete steps and porch.

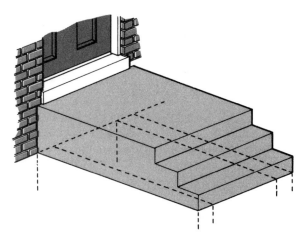

Remove forms from the finished porch and steps and allow to dry.

garden walks, perimeter walks around swimming pools, and driveways. The charm and appeal of this type of concrete is that the stone or aggregate in the mix is exposed and presents a pleasing appearance. It offers an economical, long-lasting surface.

There are two basic ways to achieve this. One is to order ready-mixed exposed aggregate concrete from the concrete company. They use a creek or pea gravel in the mix and sometimes you can specify the size stone you would like to have. I would stick to a medium size, as they seem to stay imbedded better in the finish surface. Colored chips of stone are also available from most companies, but do cost more. The stone is thoroughly mixed throughout the concrete and takes the place of regular crushed stone.

Pour the same as regular concrete, making sure you float it well, covering all of the aggregate or stone in the mix with concrete paste. Don't overfloat or you will bury the stones too far down in the mix.

After the stones are thoroughly imbedded and as soon as the concrete will support the weight of a person on a wooden kneeboard or plank without sinking in, you can begin to expose the aggregate or stones. Gently brush across the surface with a broom or medium-bristle brush, depending on the texture you want to achieve. At the same time, apply a fine spray of water from the hose nozzle onto the area. Be careful not to dislodge the aggregate or wash it out too deep or it will work loose. A little practice with this technique and you will soon learn how easy it is to master. The surface can be fairly smooth with only a little of the aggregates showing or more rustic or bold by adjusting the spray on the hose to be more forceful and using a stiff-bristle brush. If you have the misfortune to dislodge a few of the gravel or stones, have a little portland cement paste mixed up and press the stone back into the surface. However, this is not the best practice, so be careful to not dislodge the stone from the surface. How soon you begin to do this procedure will depend on drying time and weather.

Some of the rental tool companies have a special exposed aggregate broom that attaches to a hose. This really works great, as water sprays through a number of plastic holes directly onto the concrete surface. These brooms are about 18″ wide. If you are going to do a large driveway or patio, it would be worthwhile to rent or buy one of these. Goldblatt Tool Co., 511 Osage, P.O. Box 2334, Kansas City, Kansas 66110, has these available through their tool catalog.

The second method used to obtain the exposed aggregate finish is known as "seeded aggregate." This is accomplished by screeding the concrete first as normally done and then scattering by hand, evenly over the entire surface, the aggregate until it is evenly distributed. A 3-pound coffee can works well for small areas.

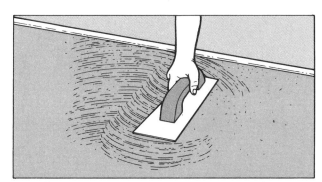

Floating the concrete and exposing the aggregate with spray of water and broom.

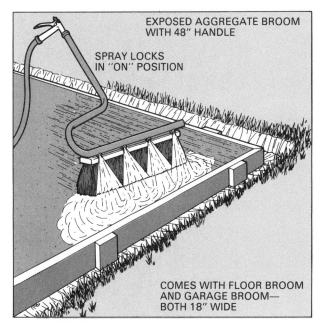

An exposed aggregate broom attaches to a hose and allows you easy access to a water supply while at the same time provides a broom for brushing the concrete surface.

Next, imbed the aggregate into the surface using a piece of wood and tapping it down flush with the surface. After this has been completed, using a circular motion, float the entire area with a wooden float until all of the aggregates have been covered with cement paste to fill in all voids. Let the concrete set until it has dried enough for the next step. This will depend on the temperature and drying time.

As previously described for ready-mix aggregate, use a fine spray from the hose and a broom to expose the aggregate. Before committing yourself to a large area, always start at one corner or edge and work on a small test area. If this works well, then go ahead with the rest of the exposing operation.

Exposed aggregate should be cured thoroughly. Exercise care that any method of curing used does not stain the surface, such as placing straw or some type of building paper. Let the sunshine and normal air dry the finish. If a slight cement stain or haze remains on some of the surface of the stones, don't worry about it, as it can be removed later by using a solution of 1 part muriatic acid to 10 parts water.

Thus far I have been describing placing exposed aggregate in larger areas such as patios or walks.

One flexible method of utilizing exposed aggregate is to make wood forms and pour individual blocks. They are not only easier to work with, but can be done over a period of time rather than all at once.

I built about eight forms of wood 2×4s, cutting 45° angles on the corners so they fit neatly and measured 20″×20″ inside. The corners were joined together with some old hinges so the forms would come apart easily, thereby allowing the concrete blocks to be removed without chipping. On one corner use a hasp instead of a regular hinge, fastening it by inserting a nail to hold it together during the forming process. This allows the form to be folded away from the concrete when the nail is removed from the hasp.

By building the form of 2×4s standing on the narrow edge with the widest dimension up, it will be strong enough to resist twisting during the forming. The thickness (3½″) allows space to insert metal wire for reinforcement without it working out on top of the finished surface.

To pour blocks, lay the forms out on a flat surface such as a driveway or garage floor using a sheet of plastic under the forms (plywood will also work just as well). Oil the inside of the form with motor oil so the concrete will not stick to the sides. Mix the

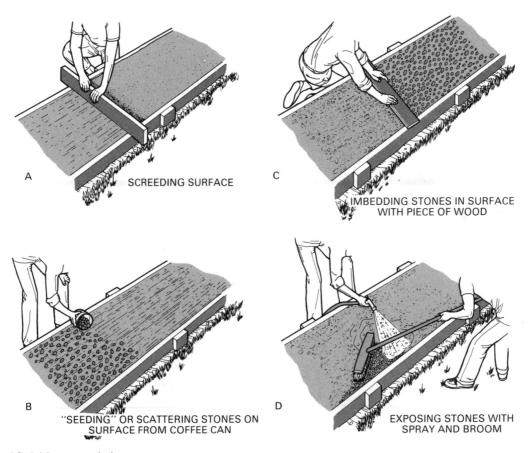

A SCREEDING SURFACE

C IMBEDDING STONES IN SURFACE
WITH PIECE OF WOOD

B "SEEDING" OR SCATTERING STONES ON
SURFACE FROM COFFEE CAN

D EXPOSING STONES WITH
SPRAY AND BROOM

Placing and finishing screeded aggregate.

concrete in a wheelbarrow in a 1:2:3 proportion previously discussed and pour in the form. Pour concrete about ½″ down from the top and sprinkle aggregate over it. Tap these down into the concrete with a piece of 2×4 and then float over them until the top is smooth. Let these set a couple of hours until they start to dry. Then the same process is used as described when using the brush and a fine spray of water to expose the aggregates.

Let the blocks in the forms until the next day, then gently remove the nail from the hasp and pull the form away. By using hinges the form does not have to be picked up; this, therefore, eliminates a possibility that the block would break. After the forms have been removed, wet the finished blocks about twice a day for several days and allow them to cure slowly, approximately one week, before using them. The beauty of making blocks this way is that the forms can be used over and over again. When enough blocks have been made to complete your patio or walk, they are ready to be put down on a bed of stone screenings and set into position using a heavy hammer.

Multiple-Grid Blocks

Flagstone is fairly expensive to use for walks or patios. A more economical substitute is to make multiple-grid blocks of concrete. They can be of exposed aggregate or troweled smooth, whichever you like. The forms can be made of regular wood 2×4s, but I prefer using 1″×4″ wood strips as the joints between them will be smaller, making for a closer fitting design.

Using a walk for an example, excavate the walk area about 5″ in depth and tamp firmly to provide a good stable surface. Fill in with about 1″ of stone screenings and tamp evenly. Make several wood forms with dividers nailed in place to suit the pattern of various blocks you desire.

You have two different routes you can follow. You can place the wood forms in the walk and pour them in place, or you can put the forms on a surface covered with plastic and then fill them with concrete and when cured lay the blocks in the walk individually. I recommend in this case pouring them in place.

A metal cabinet pull screwed onto the top of each end or a wood block will serve as a handhold making it easier to remove the form.

After making the forms and setting them in place in the walk, oil all of the inside edges so the concrete will not stick when it is time to remove the forms.

Mix a batch of concrete in a wheelbarrow and fill the forms. Trowel smooth or finish with exposed aggregate as desired. Remove the forms when the concrete has set enough and be careful not to chip the edges. This may be the next day or if the weather is nice and sunny, the same day. You will have to experiment to determine this.

When removing the forms, tap around the edges first and gently lift up using the handles. Be sure to scrub the wood forms off with a brush and water so they will be ready for use again.

Let the concrete cure about 24 hours before filling in between and around the edges with earth. One technique I really liked was where the handyman sifted fine earth between each of the joints and then planted a fine bent grass, such as found on golf courses. The result was a pleasing appearance of simulated flagstone with colorful green grass joints. If you don't care to do this, sand can be swept into the open joints as an option. However, don't fill in the joints between the multiple grid blocks with any type of mortar that will harden, as it will only

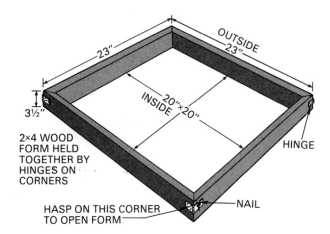

A 2×4 wood form held together by hinges on corners.

Completed patio of exposed aggregate blocks.

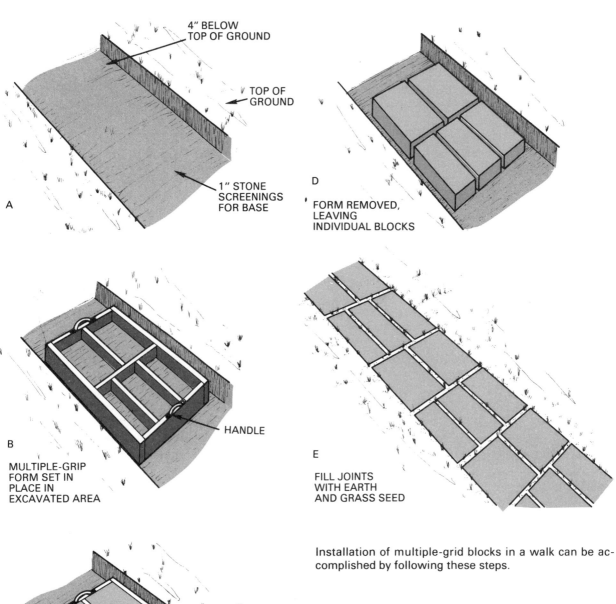

A

4" BELOW TOP OF GROUND

TOP OF GROUND

1" STONE SCREENINGS FOR BASE

B

HANDLE

MULTIPLE-GRIP FORM SET IN PLACE IN EXCAVATED AREA

D

FORM REMOVED, LEAVING INDIVIDUAL BLOCKS

E

FILL JOINTS WITH EARTH AND GRASS SEED

Installation of multiple-grid blocks in a walk can be accomplished by following these steps.

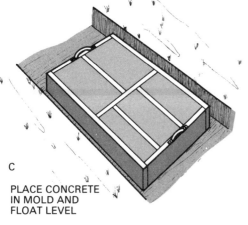

C

PLACE CONCRETE IN MOLD AND FLOAT LEVEL

result in cracking as the blocks shift when walked upon.

Making Concrete Cap Stones (Coping) for Masonry Walls

When the last course of brick, block, or stone is laid on a wall, some type of solid cap has to be built to prevent moisture from penetrating. One good solution is to form and pour a concrete cap. This cap should be laid in mortar and project about 1" on all sides and ends to allow water to drain. Instead of forming the cap level on top, I like to build the form on a slight angle to promote drainage.

Start by building a form as shown in the following illustration. I recommend using either ¾" plywood or 1" boards such as white pine or spruce for the forms as they are easier to hold and brace straight.

Make the form 2" wider than the wall, as this will allow the 1" projection on both sides. Cut the height of the board on the back side of the wall 4" high and the front ones 3½". The side pieces will taper from 4" to 3½". Nail these together with a coated nail so they will hold firmly. A 4d or 1½" common nail will be the right size. I wouldn't recommend making the length of the form longer than 3' because of the weight.

After the forms are nailed together, place them on a sheet of plywood or plastic laid on a hard, flat surface. Add a few strips nailed across the top of the form from front to back in several places to keep the form from bowing when the concrete is placed in it and keep the size true. This is shown in the illustration. Two ¼" or ⅜" steel reinforcement rods will work great for strengthening the concrete. If these are not available, cut and lay regular concrete reinforcement wire in the forms, being careful not to touch the ends or sides of the form anywhere as you would not want them to show through the finished cap when the forms are removed.

Oil the inside edges of the forms and mix a batch of concrete in a 1:2:3 proportion. Shovel or place the concrete in the forms, working it down well along the edges by slicing the bricklayer's trowel up and down. You can also tap along the edges and accomplish the same thing. This should fill any voids and make a smooth finish. Let the concrete cure at least two days before removing from the forms. They are then laid in a mortar bed on the wall, completing the job.

Casting Using the Ground as a Mold

You can make very simple stepping stones of concrete by digging out irregular holes in the ground and pouring concrete in it. These look nice in a garden walkway as you can install them to fit the curvature of the walk.

First, dig the area out where you want the stepping stone to be. Make it at least 8" in depth and 16" on the side. Cut a piece of plastic and line the hole

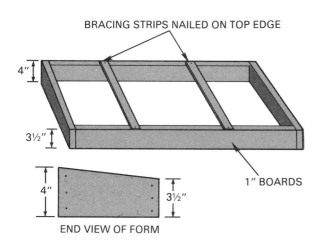

Construct a wood form such as this for the concrete cap.

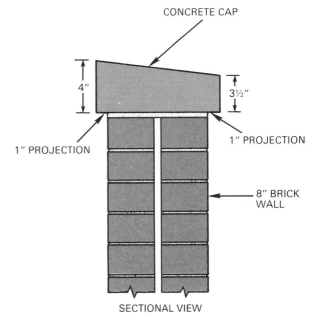

A completed concrete cap positioned on an 8" brick wall.

with it. This helps the concrete to cure slowly without the earth sucking all of the moisture out of it too quickly. Next, twist some old fence or metal wire around in the hole for reinforcement. Mix some concrete in 1:2:3 proportions and put in the hole. The top can be troweled smooth or you can press or settle some coarse aggregates in the top to be exposed. One wheelbarrow of concrete will pour a number of stepping stones. Have enough holes ready before you mix the concrete so there is no waste.

Making Concrete Look Like Flagstone

Fresh concrete surfaces can be scored or marked with a jointer to resemble flagstone or in decorative patterns. This is commonly done around swimming pools or patios, when the concrete is to act as the finished surface.

After the concrete has been screeded and troweled, it is scored or marked in random geometric patterns by using a jointer, groover, or a bent piece of ½" or ¾" diameter copper pipe about 20" long. The pipe will cause the appearance of a recessed joint in the concrete. This tool looks a lot like a regular hand-held convex jointer used for brickwork. It should be done while the concrete is still soft and pliable. I would recommend scoring it about ¼" deep.

The first jointing will leave a ragged edge. Let it dry for a short time and then carefully, with the

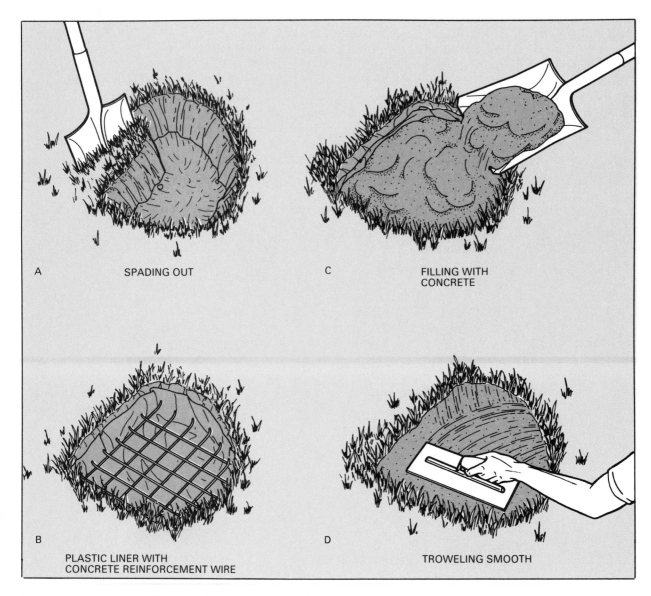

A SPADING OUT

C FILLING WITH CONCRETE

B PLASTIC LINER WITH CONCRETE REINFORCEMENT WIRE

D TROWELING SMOOTH

Forming a concrete stepping stone by using the ground as a mold.

trowel, smooth these spots out, making sure not to fill up the grooves and ruin the design. Run the tool through again one last time to make the recessed areas neat and crisp. Let it set again for about one-half hour and lightly brush with a fine-bristle brush. The result is an attractive concrete finish that looks like flagstone.

Making Circle Designs in Concrete

This is an unusual and interesting design that is very easy to accomplish. Select some tin cans of various sizes in diameter. You can make as many different size circles as you want or use the same can for all of them.

After the concrete has been troweled smooth and is still pliable, you are ready to start. It is a good idea to use the largest size first. I prefer using a 3-pound coffee can, pressing the opened end of the can into the freshly troweled concrete, and turning it slightly with your wrist to make a clean impression. Usually ¼″ to ⅜″ is deep enough to make a nice, clear indentation. After placing the large imprints where you want them, use the next size and repeat the procedure (a 1-pound coffee can is a good second

choice). Continue making circles until satisfied with the results.

Let the concrete dry enough so that it will not smear. Then with a small pointing trowel, smooth around the ragged edges where the imprints were made. Leave the concrete cure overnight and sprinkle with a fine spray from the hose the next day to promote good curing.

Wavy Broom Finish

A coarse-bristle broom has been used many times to create a nonskid surface, especially on driveways or walks that get slippery. The wavy broom finish can be used to add a little rustic effect and still achieve the nonskid results. Immediately after the first troweling has dried enough to prevent excessive smearing or sagging, use a stiff-bristle broom (floor type) and move it across the surface in a wavy motion. Alter the movement of the broom to create large wavy patterns to smaller ones. Usually only one pass across the concrete is sufficient to do this. Let it cure overnight and dampen with a light spray from the hose to ensure proper curing.

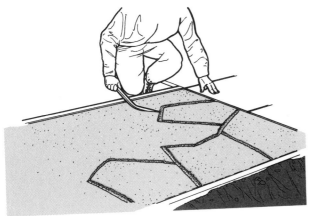

Making concrete look like flagstone involves random scoring with a bent pipe, troweling of ragged spots, and adding finishing touches with a soft-bristled paint brush.

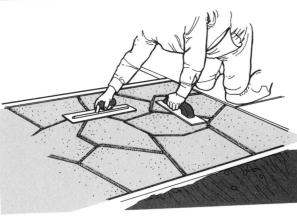

Coloring Concrete

Concrete can be colored by using any of the following three methods: adding color pigments to the wet concrete when it is mixed, using a dust or dry-shake color pigments on the surface and troweling in, or using stains or paints after the concrete has dried.

When talking to different contractors who have tried all three methods, most agree the most effective and economical is the dry-shake. This is done by applying a dry-shake coloring material available from various reliable manufacturers ready for use. Check with your local building or concrete supplier, as they should be able to supply the name of a product and the colors available in your area.

The basic ingredients are mineral oxide pigments. White portland cement can be substituted in lieu of regular portland cement (which is gray in color) to obtain a brighter color. Your building supplier also stocks white portland cement.

After the concrete has been screeded and floated as usual, the free water that comes to the top should be left to evaporate. The surface should then be refloated, working up the cement cream or paste to the top. A hand magnesium or aluminum float will give the best results.

Immediately following this floating operation, the dry-shake material is evenly distributed by hand over the surface. If too much color is applied in one spot, a nonuniform color can result, so be careful to apply evenly. The first application of the dry-shake material should use about two-thirds of the total amount specified according to the directions on the package. In a few minutes this dry material will absorb most of the moisture from the concrete and then should be floated again. Now sprinkle the rest of the dry-shake material on the surface evenly and refloat again until it is worked well into the surface. If there are any rounded edges or expansion joints, retool them at this time to obtain a smooth surface.

Steel troweling should follow the floating process as soon as it sets enough so it won't smear. Allow a lapse of time to set, and finish troweling a second time to improve the texture and smoothness.

If desired, a soft bristly brush can be used over the surface to produce a slightly roughened finish and provide a nonskid surface.

Making circles in a concrete surface with a round can.

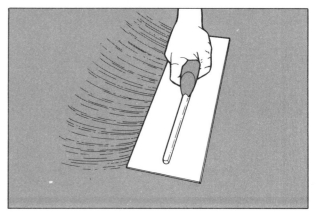

To obtain a wavy broom finish, trowel the surface first; then move a stiff-bristle brush across the surface.

Care should be taken especially with colored concrete to cure it correctly. Periodically dampen the surface with a fine mist of water as it cures to help prevent drying out too fast. About every 12 hours would be a good guide to follow. Temperatures and weather conditions have a lot to do with curing any concrete. Water should never be sprayed on any concrete if there is freezing weather or if it sets overnight when the temperature is supposed to go below freezing.

There are also other types of coloring materials available from some manufacturers. Be careful to avoid anything that would stain the concrete until it has cured, which would be at least 48 hours. Foot traffic or moving any objects across it that would mar the surface should not be done during this period. Although colored concrete can be poured successfully, it is well to remember that over a period of time it is possible there may be some fading or spotting. The dry-shake method seems to be the more long lasting.

COMMON DEFECTS THAT OCCUR IN THE SURFACE OF CONCRETE

Following is a description of some of the most troublesome defects that can occur on the surface or finish of concrete with some preventive measures and solutions.

Scaling

Scaling is when the top or surface of hardened concrete breaks away to a depth of about 1/16" to 1/4" and flakes off. This as a rule occurs in the early aging of the concrete and is caused by freezing and thawing of the surface of freshly placed concrete or faulty workmanship.

If freezing and thawing of the surface is the cause, this can be prevented by using air entrained concrete. Maintain the temperature of the fresh concrete above 50°F. for at least five days. If inside, use heaters and if outside, cover with straw or build a plastic tent around with some type of heat if below freezing. If this cannot be accomplished, do not pour concrete in freezing weather.

If faulty workmanship is the cause, it may be because the finishing operations (such as screeding, floating, or troweling) were done while the free water was still on the surface. Mixing excess water into the top of the concrete will cause a separation of the fine aggregates such as sand and cement and may cause scaling. Many times it will also cause a fine layer of cement, clay, and silt to rise to the surface. This results in a weak finish and will probably scale or flake off.

No finishing operation should be done while free water is present on the surface. Let the water evaporate or be forced to dry out by using fans or blower type heaters. Heaters of this type are available from rental stores at a reasonable cost.

Crazing

Crazing is a common problem in placing concrete. Basically, it is the formation of many fine hairline cracks on the surface of new concrete and is usually due to shrinkage. These cracks form a pattern that looks similar to crushed eggshells.

One of the causes of this problem is rapid surface drying, caused by high air temperatures, hot sun, or drying winds. This can be remedied by spraying a fine mist of water on the concrete after it has set enough so as not to smear to slow the drying process. Also, burlap, canvas, or sheets of plastic can be used to cover the concrete. Straw is also an inexpensive covering.

Another cause of crazing is premature floating and troweling before the concrete is ready and also finishing when there is an excess of water on the surface and the concrete is still soft. This brings the sand up on top of the surface, which results in shrinkage and crazing as it dries. To prevent this, don't start floating or troweling until the moisture has evaporated from the surface and the concrete has started its initial set. Don't over-float or trowel when the concrete is still soft; give it a chance to set a little and be stiff enough that it will smooth out to a reasonable level. It is better to retrowel to achieve a smooth finish a second or even a third time, as it just can't be done in one operation.

Crazing also can be caused by overuse of screeding, straight edging, or floating that brings an excess of cement paste to the surface. To prevent this, use these tools sparingly because excess cement on the surface tends to cause a lot of shrinkage and delays the time for proper finish troweling.

Dusting

The appearance of dry, powdery material on the top or surface of a newly hardened concrete slab is called dusting. It is one of the most frequent problems encountered in pouring concrete.

An excess of clay or silt in the sand or stones used in the concrete is one of the causes for this problem. Use only clean, well graded aggregates in the mix and stockpile on top of a protective covering rather than directly on the ground. A sheet of plastic will usually give good protection.

Condensation appearing on the surface before floating and troweling have been completed is another cause. This usually occurs in the spring and fall when materials get cold due to low night temperatures and the fact that nothing has been done to keep any heat on the concrete. A good example is when concrete floors are poured in a basement where there is no heat. The temperature rises slowly and the concrete sets much slower. Moisture will also condense on the surface if the air temperature goes up quickly during the day and the humidity is high. Floating or troweling this excess water into the slab can cause dusting after it cures. To prevent this problem, provide some type of heat if the weather is cold, either by using blowers or warm air from fans. If these are not available, open windows or doors to increase circulation of the air over the slab. Again, when there is a lot of free water still standing on the surface, do not float or trowel. Never use dry portland cement sprinkled on the surface to speed up the drying process, as it does not unite with the bulk of the concrete and will only dust off later. Over-troweling once the surface has set will only cause a separation of the fines (sand and aggregate) and result in dusting after the curing process.

Letting concrete set too long before finish-troweling and sprinkling water on it to work up a paste or cream can cause dusting also. This is one of the most frequent problems for a nonprofessional when finishing concrete. Once you have floated concrete, never leave it until it has been troweled.

There is no way of predicting how long it will take before the surface is ready, but be there and keep testing it with the trowel blade until it is at the right stiffness. On a nice warm, sunny day it will dry more quickly while on an overcast, colder day will take much longer. A good rule to follow to decide when to finish troweling is to press your thumb in the surface. If it leaves an impression, you had better start troweling.

Poor or inadequate curing techniques will also cause dusting. Be sure to allow an ample amount of time for the concrete to cure without rushing it. Foot traffic or parking a vehicle on it before it is at least a week old is asking for trouble. Even though the concrete looks hard, well cured concrete takes at least 28 days under normal conditions. In the summer or hot weather, remember to keep the finish dampened and in cold weather don't let the surface get below freezing. The best results will always be obtained in warmer weather; so, if possible try to schedule this part of your project when conditions are good.

I realize that it is not always possible to have all conditions perfect, but your concrete problems will be greatly diminished if you try to work with the weather. It takes planning!

34
STONE MASONRY

This section of the book contains information on sources of stone, types of stone masonry, tools needed, recommendations for stone mortar mixes, how to estimate stone and mortar you will need, good work practices, and laying stone.

Of all of the different types of masonry work, stone masonry is to me the most enduring, beautiful, and satisfying. Stone has been used for thousands of years as a building material and is still one of the most desirable. Stone masonry is something that the homeowner or handyperson can do and be very creative and successful.

Stones are relatively easy to work with, because you do not have to follow set patterns or bonds. They do not have to be laid perfectly plumb or level, but more or less as you pick them up from the pile and decide where to put them in the wall. You can be as flexible as you want, though you should follow some basic rules that will be pointed out in this section.

It is true that stonework does take patience and a great deal of time, but the effort you put forth will be rewarded with a finished project that will be the envy of your friends and give you a great deal of pride.

A few words of advice are in order before starting.

Stones have sharp edges, so be wary of cutting your hands, legs, etc., in handling them. I also strongly suggest you wear a pair of goggles or safety glasses with side shields when cutting stone, as the pieces really fly. A tough pair of jeans and a good heavy shirt or light jacket will help protect you. Gloves are a must! I like leather or leather-palm ones, which most farm supply stores stock. It's too late to put on the gloves once you've suffered a nasty cut.

Sneakers are taboo for working around stone. I suggest a fairly stout pair of leather shoes to offer some protection if you drop a stone on your foot.

If you will be laying heavy stones, don't attempt to lift with the stone held out away from your body, but try to let your legs take the weight rather than your back. When you have a large heavy stone to work into the wall, get some help or try sliding it into position on a board.

This is one type of masonry work in which you can easily get hurt. The safe way is always the best way, even if it takes a little longer. But have fun, and enjoy it! Laying stone is one of the best therapies I know of, after a long hard week working in a different type of job.

WHERE TO OBTAIN STONE

Stones are everywhere! The type of stone you select will determine where you look for it. Spend a little time and in all probability you can get your stone free, not counting gasoline hauling it home. Good sources of building stone are old fence rows, rock outcroppings in fields or mountains, along streams, and old deserted buildings such as barns that are falling down. You can, of course, buy stone from building suppliers or from a quarry. This is rather expensive . . . but it depends on the type of stone you are looking for.

A word of caution: Be sure to get permission before taking any stone from someone else's property, or you may wind up in court. As a rule, most farmers and landowners are more than glad to get rid of stone for a small cost, or perhaps you can swap a little exchange of labor.

TYPES OF STONE MASONRY

It will be helpful if you know a little about the two general classifications of stone masonry work: rubble and ashlar.

Rubble can be of two types: rough and roughly squared. Rough rubblestone work is built of irregular stones of varying sizes and shapes as they come from the stone pile. The edges are usually left as they are, with little or no cutting or dressing with the hammer and chisel. The mortar joints are usually rubbed out with a piece of wood or merely raked out and brushed. This presents a very natural rustic appearance.

Roughly squared rubblestone work takes more time, but has more of a cut, dressed appearance. The stones are cut on the edges with the top of the stone laid in a somewhat level position, unlike rough rubble. There is no specific repeating bond pattern, but stones should lap over those beneath with few or no vertical joints in line. Stone that lies in layers in the ground lends itself best to this type of stonework. A rakeout joint highlights a roughly cut stone, as it shows the edges well.

Ashlar stone masonry is the type you see in churches, banks, university buildings and other

The remains of an old stone barn—an excellent place to obtain stone. But make sure you get the landowner's permission! You don't want to end up splitting stone at the county jail.

In rough rubblestone work, very little cutting or dressing is done; the stones are used as they come from the pile. The terraced rubblestone wall is pleasingly rustic.

structures that are intended to have a somewhat formal appearance. The stones are squared on all edges and are usually larger than a brick, and all of the mortar joints are about the same size. The bond can be in a repeated pattern, or varied with two or three stones laid against a larger one. This forms a neat but irregular pattern. Ashlar stonework is often used when building the fronts of fireplaces and buildings of sandstone, because the sandstone cuts easily and comes from the quarry mostly square, with only the lengths being cut. It lies in layers in the rock deposits and is readily taken with a minimum of effort.

TOOLS FOR STONEWORK

Although the professional stonemason has a large assortment of hand tools, your list should be relatively simple. Many of the tools listed in the chapter on brick will do for stonework. However, you are going to need a couple of heavy hammers, chisels, and jointing tools. A sledgehammer is also a must for breaking up large stones and knocking off

edges. Sledgehammers vary from 10 to 16 pounds in weight. It is helpful if the sledgehammer has an ax-tapered edge rather than two square ends.

In addition to the sledgehammer, there is a hammer made especially for cutting stone. It looks like a sledgehammer but is smaller. It weighs about 4 pounds and works great for trimming or cutting stone. You should also have a square-head mash hammer; a 2-pound or 3-pound size is a good choice. This hammer is used to pound on the chisel to make a cut, and it also works excellently to settle a stone in the mortar bed. I suspect that it may have gotten its name from the many fingers it has mashed—so be careful to hit the chisel head and not your fingers. A regular brick hammer is fine for light cutting, but select one of the heavier hammers for doing any amount of stonework. All of these stone tools are available from Goldblatt Tool Co., which was mentioned in the bricklaying chapter.

You will need a few chisels to cut and dress stone. If you want to invest the money, I recommend you buy three different ones, as they will serve all your

Ashlar stonework employs stones squared on all edges, and the mortar joints are uniform and tooled. The pattern can be either regular or irregular.

In roughly squared rubblestone work, the stones are cut to be more or less square and are laid as level as possible, but there is no overall pattern.

A sledgehammer for stone (note the ax-tapered head) and a mash hammer, which is like a miniature sledgehammer.

needs. The most frequently used one is a plain chisel about 8″ long. It is used for general cutting of stone for length and height. The second one is a pitching chisel. This is a much heavier, broader chisel and is used for facing stone, as in cutting a rock-faced edge. The third is a point chisel, which is drawn to a single point on the end. Its purpose is to cut off bumps or rounded places to make the face of the stone straight. All of the pressure or force of the point chisel is applied to a very small area, which allows cutting off bumps or knobs without chipping or cracking the rest of the stone.

If you decide not to buy all the different tools mentioned and go with only one hammer and one chisel, I would select the plain chisel and probably the sledgehammer. But all the tools mentioned are well worth buying if you have any amount of work to do. They should be available from most building suppliers who handle mason's tools.

The regular brick set chisel is not suitable for cutting hard stone, because of the thinness of the blade and the way it is tempered. You will only ruin it if you try to use it for stonework.

The tooling of the mortar joints between the stones can make a world of difference in the overall appearance of the finished wall. Selection of the joint finish depends on the type of stone being laid. For example, a rubblestone wall with irregular stones in it looks good with a rakeout joint formed with a wood stick or a piece of broom handle, then brushed clean. Another popular method is to smooth out the joint with a pointing trowel and brush it lightly. This gives you a full or flush joint. There is also what is known as a "rolling bead" joint. This is formed with a concave stone beading tool and causes the mortar to roll out in a bead or convex shape. The choice depends on the taste of the person laying the stone. What one person likes, another doesn't.

Although you can use a pointing trowel to form a flat or flush smooth joint, it is done best with a "steel slicker." I like it better than the pointing trowel because the slicker blade forms the same size impression throughout the joint without variations. The pointing trowel also has a tendency to smear the edges of the stone.

A double-ended rolling bead jointer. The ⅜″ size is likely to be the most useful, but various sizes are available.

Examples of a plain chisel, a much heavier pitching chisel, and a point chisel.

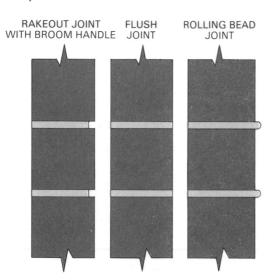

Three joint finishes used in stonework.

Slickers for forming smooth flush joints are available either double-ended or with a wood handle. They do a neater job than a pointing trowel.

All of the joint finishes mentioned should be brushed lightly after they have set enough not to smear.

MORTAR FOR LAYING STONE

Over the years I have found that a mix of portland cement, hydrated lime, and sand is best for stonework. I do not recommend using masonry cement mortar for stone, as it will not bind to the stone nearly as well as portland cement mixes.

My favorite mix for an average stone mortar is as follows: 1 part Type 1 portland cement to 1 part hydrated lime to 6 parts sand, with enough water that the mortar will support the weight of the stone without settling or smearing.

A richer mix for stonework that will be subjected to a lot of moisture or is below ground level is as follows: 1 part Type 1 portland cement to ½ part hydrated lime to 3 parts sand.

A good mortar for stonework has to be not only strong but elastic and workable. The addition of lime to the mix gives it these qualities, and more important, it creates an excellent bond with the stone.

Because of the hardness and density of stone, you will have to experiment a little to determine what is the best stiffness of mortar for the particular stone you are laying. There is no single rule for all mortars. Just remember that the mortar must support the stone without sinking or bleeding. It is also a good idea to select a sand that is somewhat close to the color of the stone. Don't buy a very light sand if the stone is dark, unless you want a whitish mortar. As for mortar for brickwork, buy clean, washed sand whenever possible. Another useful tip is not to mix any more mortar than you can use in one hour. This eliminates adding water lost to evaporation and weakening the mix.

ESTIMATING STONE AND MORTAR

Estimating stone can be tricky! One can be very careful in figuring the exact amount, but in the case of rubblestone, a considerable amount of the wall area will be occupied by mortar because of the many irregularities in the shapes of the stones.

Rubblestone is sold by weight. The simplest method of estimating how much stone you need is to determine the amount of cubic feet in the wall or project and then find out how much 1 cubic foot of the stone you are going to use weighs. It is then a simple matter of multiplication to arrive at the correct figure. This is the way most suppliers that I have dealt with over the years sell stone.

A method I have used is to select a stone from the pile or at the quarry that is approximately 1 cubic foot (12"×12"×12") and weigh it on a freight scale. Record this weight and when buying your stone, check the total weight bought and divide by your figure of weight per cubic foot. This will tell you, as a rule, if you are getting your money's worth. Almost all stone dealers give the customer the benefit of the doubt when making up the bill. I have also found that on the average 1 cubic foot of stone will weigh about 125 pounds, but you cannot always go by this. It is a good idea to check the stone you are buying.

There is an older method of estimating stone, still in use in some parts of the country, based on the "perch." A perch of stone is considered to be about 16'6" long, 1' high, and 1'6" thick. This equals 24¾ cubic feet. Different localities have different versions of a perch measurement or ways to compute it. In a number of states west of the Mississippi River, rubblestone work is figured by a perch containing 16½ cubic feet. My advice is to find out beforehand how the person is selling you the stone and go from there.

A load of squared stone in random lengths and heights, as sold by masonry suppliers.

So far, I have assumed you want to buy the stone from someone by weight. Many farmers or landowners will simply price it to you by the pickup load, with no reference to weight. This, of course, is the best deal!

Last, if the stones are free, then there is no need to bother with the estimating procedure. Simply get a load and lay it in the wall, and from the area it occupies, you can determine how much more will be needed.

If you want a squared ashlar stone for a fireplace front, porch, etc., many building suppliers stock it, usually in a variety of lengths and heights but in a standard 4″ or 5″ width. You give the supplier the number of square feet of wall surface and he will tell you how many square feet you need, based again on a square foot, not a cubic foot, of standard weight. It is a very accurate method of figuring and there is little or no waste.

One more tip about buying or obtaining stone. If you are going to the quarry or some farmer's land, don't load up with all large or all small ones. Try to pick up a variety so they can be worked in the wall more easily and will form an interesting pattern. Be a little selective and it will pay off in the long run. Most people you buy from are only interested in filling your truck as fast as possible and getting paid. Work out this agreement beforehand and there will be no arguments or misunderstandings.

The stones should also blend somewhat in color and texture and be compatible with one another. A good rule to remember is that the face area of the biggest stone should be approximately six times the face area of the smallest.

Estimating Mortar for Rubblestone Work

Because of the many different shapes and voids in rubblestone work, it takes a great deal of mortar. In fact, a third of the wall area could be mortar. There is no exact method of estimating mortar for rubblestone work. What most masonry contractors who lay stone do is to depend on past job records and allow more than really needed to cover any shortcomings. If you go by the book, you need about 8 cubic feet of mortar for 1 cubic yard of rubblestone.

My recommendation is to buy a couple of bags of portland cement and hydrated lime and some sand, then lay up a section of the wall. Note how far the mortar goes and use this as a guide. You don't want to be stuck with a lot of excess materials at the end of the project. Laying a sample section is the best way to go, because it will determine the quantity of mortar you will be using. Refer to the mortar mixes given earlier in this chapter when you are ready to start.

How to Figure Flagstone

There are two basic types of flagstone used for masonry, natural random rubble and a smooth quarried stone. Flat fieldstones and random stones from the mountains, along a stream, etc., constitute random rubble flagstone. This is, of course, inexpensive; it can be obtained cheaply or for the taking.

The best method of figuring how much flagstone you will need is to weigh a piece that is approximately 12″×12″ and record this weight. It is a good idea to try to stick with a stone thickness around 1″ to 1½″ as such stones are easier to work with. Next, figure the square footage of the area to be covered, and then multiply the weight of 1 square foot by the number of square feet in the project. This should give you a rough idea of how much stone you need. If you are obtaining your stone free or inexpensively from someone, it will probably be priced by the pickup load and you will not need to get involved in close estimating.

Selecting Flagstone

Many building suppliers sell a slate flagstone that is very popular. Generally this slate comes from the Middle Atlantic and New England, mostly Vermont, and thus the name Vermont slate. There is also a considerable amount of flagstone sold from New York State, although it is not a slate.

Slate flagstone is taken from quarries and lies in layers of approximately even thickness. This is nice as it is easy to work with when leveling in the mortar bed. It comes naturally in beautiful colors such as red, green, blue, purple, and a brownish tone. The edges are straight as they are sawn at the quarry. However, the stones are not square but more or less rectangular. When you're building a walk, porch, patio, etc., they are laid much the same as when fitting a puzzle together.

Slate flagstone initially is more expensive than other types because of the sawn edge and shipping costs. However, it is still cheap for what you get, because there is little waste and it is very easy to work with. It has long been a favorite of builders, homeowners, and the handyperson.

The thickness of slate flagstone is fairly consistent, varying from about ¾″ to 1½″. It is usually sold by weight, allowing 10 pounds for every square foot of area. This is based on a 1″ average stone thickness and works out very close. Don't allow any extra for mortar joints but figure the area as solid stone; this small difference will usually compensate for any breakage or problems encountered.

Mortar for Flagstone

I strongly recommend using only a mortar of portland cement, hydrated lime, and sand. I also recommend a richer (stronger) mix, as it will resist moisture and cracking in flat surfaces where the stone will be laid. My favorite mix is as follows: 1 part portland cement to ½ part hydrated lime to 3 parts sand.

Allowing an average mortar bed joint of about 1″, it takes approximately 1 cubic foot of mortar for every 12 square feet of flagstone. This figure will, of course, vary a little if your concrete slab is not level.

HOW TO LAY STONE

The best practice to follow before actually laying any rubblestone in mortar is to select a flat spot and spread out the stones so that you can see what you have to work with. Many times, after some practice, you can look at the pile and pick a stone for the wall that does not need any cutting. Don't forget to wear good heavy shoes; the sharp edges can inflict a severe cut to the foot or leg.

Assuming you will be laying the stone wall in mortar, it is necessary to excavate below the existing frost line in your area and pour a concrete footing. As a rule the footing should be approxi-

mately 8″ thick and about 4″ wider than the wall to be built on it. If you need a large amount of concrete, you can order it by the cubic yard from a local concrete company. If you need less, you can mix your own by using a utility mixer, to a proportion of 1 part portland cement to 2 parts sand to 4 parts stone. This is what is called a 5-bag mix to the cubic yard and is recommended for footers.

After the footing has been poured and has set for at least one day, establish some level points on both ends of the wall so you know where you are going for height. This can be done by using a good straight board with a level laid on top of it, or for greater distances a hose that is transparent can be filled with water and let down about 2″ from the top to act as a water level. To use the garden hose as a level, hold it up on both ends of the wall with the help of another person and mark the water level on a stake. This is an excellent method of obtaining a level point, as water will always lie level.

Unless you have plenty of stone available, I suggest laying a couple of courses of concrete blocks below the grade line in the ditch on the footings to

Slate flagstone, sold by many building suppliers, is not cheap, but it is very easy to work with and there is little waste. It comes in several handsome colors.

A typical utility mixer like this is suitable for mixing concrete or mortar.

Before you start building with rubblestone, spread your stones out so that you can see what you have. Then you can pick an appropriate stone for each spot as you come to it and do a minimum of cutting.

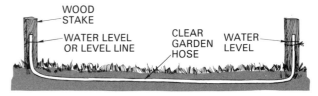

How to use a hose as a level. If you don't have a transparent hose, buy two short lengths of clear plastic tubing and insert them in the ends of a regular hose.

save your stone. Use a mortar mix of 1 part portland cement to 1 part hydrated lime to 6 parts sand, and keep it on the stiff side so that the stone doesn't sink in too far.

When starting the stonework, lay in mortar a good-sized stone on each end of the wall line and attach a line from one end to the other. Don't be worried about keeping the wall level, as stonework is not supposed to be in level courses like brickwork. The same is true for the plumbing of the wall. Stones are laid more or less from bump to bump and will be plumb overall. The natural differences in the faces of the stones are what makes them attractive.

As the work progresses, take a good stiff brush and brush off any dirt or vegetation that is on the bed side so that the mortar will bond to the stone well.

Generally, the largest or heaviest stones should be in the lower part of the wall. You can lay some big stones here and there as the wall is built, but the larger ones will be more stable in the lower section and will certainly be easier to work into position there. It also is advisable to lay a flat stone that is approximately the width of the wall at intervals, to bond the entire wall together. These are known as bond stones. I generally try to pick some flat ones that are not too high to do this job.

Try to hold the mortar joint width or thicknesses to no more than 2″ as much as possible. Large mortar joints have a tendency to shrink away from the stone and crack around the edges.

Use your sledgehammer to break up large stones that are too heavy to handle, and cut off angular projections or bumps that may cause problems.

Try to lap one stone over the other to form a bond and make the wall look natural. Occasionally lay one larger stone with two smaller ones against it. This will not only make a strong wall but create an interesting pattern.

A little trick I have learned over the years is to use the handle of the stone hammer as a gauge or rule,

Laying out the first course of stone. Use large stones at each end and attach a line to keep the course straight, but don't try to keep it level.

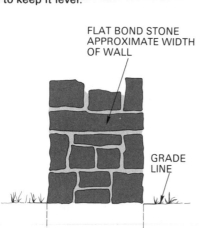

FLAT BOND STONE
APPROXIMATE WIDTH
OF WALL

GRADE LINE

Every so often, use a stone the whole width of the wall to tie it together.

Although rubblestone rarely needs dressing, it will help if you cut off the most troublesome bumps and projections.

It helps to dry-fit stones before laying them in mortar. Otherwise you'll occasionally have to take up a mortared stone, making a mess.

Lapping stones as much as possible will make the wall stronger and more natural-looking. Try for attractive combinations of large and small stones.

in place of taking the rule from my pocket and measuring all of the time. For stones smaller than the hammer, just hold your finger at that point and then try it on the stone you want to lay in the wall.

There are going to be times when you need to wedge or chink stones in order to prevent them from falling out or keep them where you want them. I suggest you find an angular chip or sliver of stone shaped like an arrowhead and drive it snugly into place. Don't drive it in so hard that it will be difficult to remove when you do the point up. Make sure that you point up fully with mortar around the wedge, so that when the wedge is removed later the stones stays in place. It doesn't take long for the mortar to set up enough to remove the wedge.

Cutting Stone

There are several methods of cutting stone, none of them difficult. The simplest is chipping or cutting a fine edge with the regular brick hammer. This is done by a light pecking action and will work for thin edges.

I have always found it awkward to cut stone that is lying on the ground. You can build a simple, very useful stone-cutting bench from some old 2×4s and a piece of plywood. This will allow you to mark and cut the stones more easily and with less strain on your back.

A good average height for the top of the bench is about 28". Use either 2×4 or 2×6 lumber for the legs and top frame. Make the bench about 32" square,

A trick of the trade—using the handle of the stone hammer as a rough gauge rather than constantly taking out the rule.

Mortar each stone in fully to make the wall solid and strong. Tapping them with the mash hammer will settle them firmly.

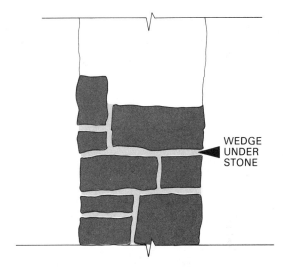

WEDGE UNDER STONE

A temporary wedge, such as a sliver of stone, can be used to hold a stone in place while the mortar sets. Pull the wedge out before you point up the joint.

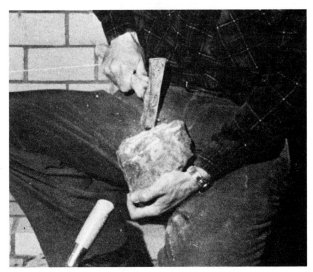

Stone can be trimmed to shape by pecking at it with a regular brick hammer.

with crossbraces to make it stable, and attach a top of plywood ⅝″ or ¾″ thick. If you use 10d nails it should hold together a long time. You are now in business.

The pitching chisel is used to cut off projecting places on the stone or make a cut face. When using it, always slant the end of the chisel slightly outward so that it will cut on a slight angle.

On occasions, there will be bumps or stubborn points of stone that you will want to cut off. For this, use the point chisel, placing the point at the very spot you want to cut. All of the force of the blow concentrates on one small area, removing the bump without cracking or damaging the rest of the face.

To split a stone, locate the grain first if possible. Usually a stone does have a grain line where it was joined together when formed in nature. Place the regular plain chisel along this grain line and strike a blow, scoring the stone. With any luck at all, this will split the stone apart. Trim off any feather edges left so that the stone will rest in the mortar securely.

Try to think ahead and save some flat stones for the top of the wall to make it level or approximately even throughout. A line stretched from one end to the other will guide you as the last course is laid.

The top of the wall should never be raked out, but pointed flat to keep water and snow from penetrating and causing deterioration of the mortar joints. It is also a good practice to let the top course project

A stone-cutting bench, well-used, made from scrap lumber and plywood. You don't absolutely need one, but it is handy to have.

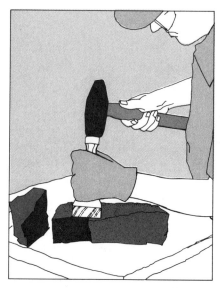

Cutting with the pitching chisel. Note the angle of the pitching chisel—slanted slightly away from the cut face. Don't miss the chisel and hit your hand.

out about 1" on each side of the wall so that water will drip off of the edges rather than run down the face of the wall.

Tooling the Mortar Joints

As previously mentioned, there are three popular joint finishes for stonework—flat, raked, and rolling bead.

To point up with a flat joint, first point up all the head and bed joints with the small pointing trowel, pushing the mortar firmly and fully into the joints. If you used full joints when bedding the stone in mortar as it was laid, it will greatly decrease the amount of pointing needed.

A flat mortar joint can be improved and made to look better by pulling a flat steel slicker tool

Removing a projection with the point chisel, which concentrates all the force on a small area.

Although you have been using the line only to keep the wall straight, not to keep the courses level, you will want the top of the wall to be of flat stones and fairly level.

Use the plain chisel to split a stone along its grain. (Some kinds of stone do not have a grain, but most kinds that are used in stone masonry do.) Trim off any feather edges.

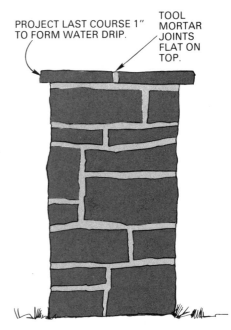

The top of the wall should be constructed to protect the wall from moisture, with flat mortar joints and a projecting water drip on each side.

through it. Tuck the mortar well into the joints as this is done. You will have to let the mortar stiffen enough so that it does not sag before running the slicker tool through the joints. If the mortar dries too quickly, wet the slicker blade to help smooth out the joint.

One of my favorite joint finishes is to rake out the mortar joints to an approximate depth of ½" with a rounded-off piece of old broom handle or dowel. Make sure that the edges of the stones are rubbed clear so that the stone will be highlighted. This should be done only after the mortar has dried enough not to sag or smear.

Brush the joints lightly with a soft-bristle brush, closing up any holes and filling out any voids that are left. A regular stove or whitewash brush works fine for this job.

Another joint finish that I really like is the rolling bead joint. The best and most effective method of tooling this joint is to go ahead and build the stone wall to the height desired, raking the mortar joint out roughly ¾" in depth and then pointing all at one time. There are bound to be some joints that will need cleaning or cutting out before the pointing process can begin. Cut with a regular mason's chisel any joints that did not get raked out before pointing. Cut on a V or angle so that the mortar will get a good bite and bond to the original joint when the

repointing is done. Brush out all of the joints to remove any remaining particles or dust left from the chiseling.

To repoint, I like a little richer mortar, as it works better and makes a smoother joint finish. I recommend using a mix of 1 part portland cement to ½ part hydrated lime to 3 parts sand. This will make a gray mortar when it dries. Some people like a white mortar joint to set off the stone. If you desire this, mix 1 part white portland cement to ½ part hydrated lime to 3 parts white sand. White sand and white portland cement are stocked by most building supply dealers who have masonry materials. Gray and white mortars are of approximately the same strength.

Now that you are ready to start pointing, get a small can of water and an old paintbrush. Moisten the mortar joints in an area that you think you can point up in about 15 minutes. Don't soak the joints, but merely dampen them with water. This is done so that the fresh mortar being applied doesn't dry out too fast and lose its strength.

To form the bead joint, pick up some of the mortar on the concave jointing tool by pressing it against the mortarboard, the same as was described for picking up mortar on the round jointer in the bricklaying chapter. Then, with a smooth steady motion, start at the top edge of the joint and

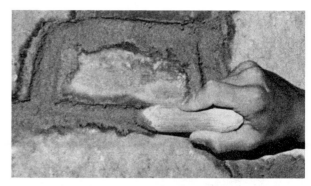

You can point up a flat joint with a small trowel. Once the mortar has stiffened somewhat you can improve the joint with a flat steel slicker tool.

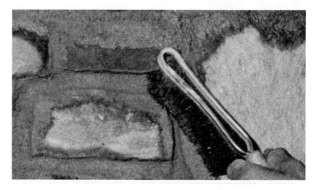

The mortar joints can be raked out with a rounded-off dowel to highlight the stone. Brushing the joints clears away fragments of mortar.

Preparing to make a rolling bead joint. The wall is built to the desired height, with the joints raked out about ¾″ and pointed. Any joints that were not raked out to the proper depth are cleaned out with the chisel. Then the joints are brushed to remove fragments and dust.

Before applying mortar for the rolling bead joint, moisten as much of the prepared raked-out joint as you can point up in about 15 minutes.

press the mortar into the jointed area, moving the tool downward or lengthwise depending on the direction you are taking. If applied correctly, the mortar will release from the jointer smoothly and form a neat beaded joint. Be careful about going over the same joint with the tool, as it will have a tendency to pull away from the rough joint.

Where two joints meet, blend in as well as possible so that the meeting is not evident. The idea is to make a smooth continuous rolled joint with no breaks or interruptions. This will take some practice before you get it down perfectly.

After the completed mortar joints have dried enough, brush lightly with a soft brush. The result will be a beautiful stone wall with a raised rolling bead joint.

If you have a problem obtaining a concave jointing tool, you can make one from a 10″ length of ¾″ copper water pipe. Saw the copper pipe in half along its length for about 3″ and file the cut smooth. Sweat or solder an elbow onto the end to form a handle. This tool does an excellent job and will make a nice smooth joint.

Laying Flagstone

Flagstone can be laid in mortar on a concrete base or dry in a bed of stone screenings, sand, or both. It is, of course, more difficult to lay them in a mortar bed, since the joints must be tooled and the flagstones must be carefully leveled. However, neither method is hard.

Making the rolling bead joint. This is slow work that requires patience and practice; it is particularly difficult to blend the beads in where two or more joints meet. However, the result is well worth the trouble—a wall with a particularly striking texture.

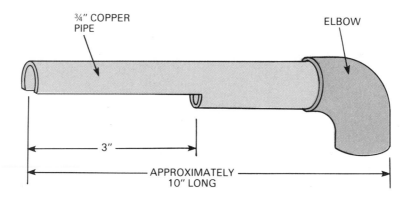

You can make your own rolling bead jointer from copper pipe. File the sawn area smooth.

Refer to the information on flagstone earlier in this chapter if you need to know how to estimate the quantity and what mortar mix to use.

Flagstones can be laid in a random square pattern or in irregular pieces that form a jigsaw-puzzle pattern. They are very popular for porches and steps and provide a surface that will never wear out. The best practice is to always lay out a border of the stone in mortar first, and then stretch a line across the borders, filling in between to the line. Lay a good straight 2×4 across the stone to tap on with the hammer to settle stones into place.

I always lay the flagstone out dry first in the area to be covered and cut them to fit, allowing about ½" to ¾" for the mortar head joints. After being satisfied that they fit, I butter or plaster mortar on the backs of the stones and lay them into the mortar bed. It has been my experience that if the backs of the stones have been plastered first, they are less likely to come loose.

To cut flagstone, first mark with a pencil or crayon on both sides where the cut is to be made.

Then lay the stone over a pipe or board so that the line is even with the pipe or board below. This will support the edges when cutting. Cut with the standard mason's chisel along the line, turning the stone over and repeating the cut on the other side. If done correctly, the stone should cut neatly along this line.

If only a small edge or point needs to be cut off, try using the head of the brick hammer and chipping lightly on the edge. The stone will usually flake off with no problems. By all means wear gloves while cutting, as the edges are very sharp. Flagstones are brittle, so don't try to take too much off at one time and work in from the edges when chipping small areas.

I have found that the pointing in of the mortar joints on top should be done as the work progresses rather than raking them out and doing them all at one time later. This is because when you point the mortar in the top joints immediately it will unite better with the mortar bed and all of it will cure together. If this is not convenient, then rake out the

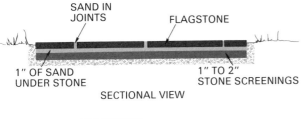

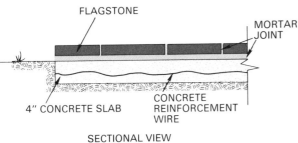

Flagstones can be laid either dry or in a mortar bed. Laying them dry is much simpler, but the mortar method is not really difficult.

Flagstones being fitted into position on a mortar bed. I like to lay out the stones dry first and cut them to fit.

Flagstone can be used in combination with other types of stone and masonry. Here it is shown capping a stone-and-block flower bed.

mortar joints on top to a depth of at least ¾" and repoint later. Use a pointing trowel to fill large areas and a slicker to form the finish joint. Brush after the mortar joints have set enough not to smear, and restrike with the slicker for a smooth joint.

If you are going to clean the flagstones, let them cure about a week, then scrub with a solution of 1 part muriatic acid to 10 parts water, and rinse well with running water from a hose with a nozzle. I consider this cleaning essential if the full color of the flagstone is to be revealed. It also removes any smears of mortar.

Laying Dry Walls

This chapter would not be complete without mentioning stone dry walls. "Dry wall" means that no mortar is used in its construction. This has several advantages. It is less expensive, it is faster to build, and you don't have the problems from freezing and thawing that are associated with using mortar. Aside from these advantages, dry walls are creative and relaxing to build.

There are a few basic recommendations in building stone dry walls.

- You don't need a concrete footing. Just excavate the earth a little to form a solid bed. This is usually about 12". It is a good idea to spread a layer of sand or stone screenings and tamp it to start off on a firm base.
- Build the dry wall much the same as a wall laid in

mortar—build both ends first, then fill in the center to a range line.

- Periodically, tie the wall together by laying some larger flat stones across. Wedge smaller ones in place to keep the others from moving or shifting.
- Don't lay stones in level courses as in brickwork, but try to create an interesting natural pattern.
- A curved wall is going to be stronger than a straight one, especially because there is no mortar present to bond the wall together.
- You can leave a small stone out here and there and pack some earth or peat moss in the pocket. Later a creeping vine or plant will grow in this pocket and create a nice effect.
- Think ahead and save some flat stones for the top of the wall. Some clay or earth can be sifted into voids or holes in the center of the wall to solidify everything.
- Start the wall wider at the bottom and taper it to the top for more strength. This angle will act as a wedge to resist movement. A good rule of thumb to remember is to batter (taper or slant) the wall about 2" for every 12" of height. This can be adjusted, of course, according to the squareness or roundness of your stones but is an average recommendation.
- If water will drain and collect against the base of the wall, it is a good idea to cut a small gutter to provide a runoff. A severe problem is best dealt with by laying some drain tiles and connecting them to a drain leader line.

Cutting flagstone. Use a pipe or board under the cut to support the stone, and cut with the mason's chisel on each side.

Dry walls are a lot of fun to build—it's not fussy or messy work, and the whole family can help.

35
MAINTENANCE AND REPAIR OF MASONRY WORK

Once you buy or build your house, as the years pass by, there will be a need for either repair or maintenance work to protect your investment. The longer you let the needed work go, the more expensive or time-consuming the repair will be.

The same is true if you are buying a building for investment purposes, which is an excellent way for the average person to invest and receive a good return on his money. If you can do your own repairs you can avoid constant calls to a contractor. The chance of a wall cracking or leaking in a new house is rather remote, but in older houses, cracks and leaks are common.

This chapter deals with some of the more frequent problems that the homeowner encounters and recommendations on how to solve them. Typical jobs are repointing mortar joints, repairing cracks, removing paint from brickwork, cleaning chimneys, parging cement mortar, and removing stains on brickwork.

Over the years I have been associated with the masonry trade, I have discovered many techniques

of repairing masonry. In this chapter I'll share some of them with you. I hope they'll enable you to do your own repairs and avoid calling in a professional.

A visual inspection is the best way of pinpointing potential problems in masonry. The two factors that cause most masonry problems are moisture entering the wall, and stress from freezing and thawing. Major structural renovations such as underpinning, tearing down of load-bearing walls, and complete demolishing of old masonry are best left to the professional contractor. If you still decide to work like this, I recommend calling in a consultant before starting the work. You will also want to check with your local building inspector or department of permits to make sure that you don't violate the law in your area.

CUTTING OUT AND REPOINTING MORTAR JOINTS

One of the most frequent types of repairs is the cutting out and repointing of mortar joints in brickwork or stonework. The cause is usually that the old mortar does not have any cement in it but is merely a lime-and-sand mix, which over the years wears out.

Many of the bricks used in older houses were soft, because they were not burned in a kiln properly, with controlled heat, as we do today. Also there was little or no quality control. Therefore, the mortar used to repoint the joints should be of approximately the same hardness as the original so it will bond to the bricks well. Using mortar of the wrong hardness is one of the most common mistakes made when repointing old brickwork.

Start your repointing job by first cutting out the mortar joints to an approximate depth of ¾" to 1". I suggest buying a plugging or joint chisel for this task. The blade of the chisel is made on an angle to help clean the mortar from the joints and to prevent binding and chipping the edges of the bricks. I recommend a medium-size chisel. Such chisels are available from building suppliers and tool catalogs.

Try to cut the mortar joints out with the chisel as square as possible, rather than in a V shape, so that the new mortar will fill in more fully. Check periodically with a rule to make sure that enough depth is cut out.

After the joints have been cut out, take the slicker tool and rake out any excess mortar or grit that still remains. Then brush out all of the joints to remove any loose mortar or sand. This is important if the new mortar is to get a good grip against the old. A

A plugging or joint chisel, used to cut out old mortar joints before repointing.

Cleaning out mortar joints. First use the plugging chisel to cut out ¾" to 1" of old mortar, checking the depth frequently. Rake out particles of mortar and grit with a slicker tool, then brush out any remaining dust. You can then spray the wall with a hose, but don't soak it.

regular garden hose with a spray nozzle works fine to flush out any remaining particles. Don't soak the joints but just remove the excess grit that is left.

If you run into some hard spots, a circular grinder can be rented to cut them out. Be especially careful, however, that the edges of the bricks are not damaged in grinding. I would use a grinder only as a last resort.

Any bad bricks that are deteriorated should be cut out and replaced with good ones. This is done by cutting the bad bricks out completely and selecting replacements that match the originals as much as possible.

Mortar for Repointing

My recommendation for a good average repointing mortar is 1 part portland cement to 2 parts mason's hydrated lime to 8 parts sand. This is a high-lime mortar and will bond excellently to the old mortar joints.

I strongly recommend prehydration of the mortar, which will greatly reduce the shrinkage of any of the joints away from the edges of the bricks. To prehydrate mortar, first mix all of the ingredients with enough water to blend them together. The mortar should be a damp unworkable mix that will retain its form when pressed in a ball in the hand.

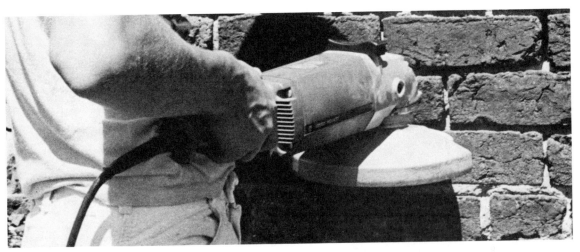

Hard spots that resist the chisel can be ground off with a circular grinder, but you must be careful not to damage the edges of the bricks.

Replacing a bad brick. Cut the old brick out completely and clean out the recess carefully. Select a new brick that matches the old ones as closely as possible, mortar it in, and fill the joints.

Let the mortar set for about 30 to 45 minutes and add enough water to make it workable. This will be a little dryer mortar than used for laying bricks in the wall. The prehydration process helps greatly to cut down on the hairline cracks that occur when pointing mortar starts to dry against the old bricks. You don't want to make a big batch for repointing, but just an amount that can be used in about 45 minutes. A little experimenting will soon determine how much you will need for that period.

Before repointing the fresh mortar, dampen the mortar joints with a brush and some water to ensure a good bond.

Pick up the mortar from the trowel with the slicker tool and press it into the joints. I recommend pointing the head joints first, then the bed joints to maintain good straight bed joints. Be sure to pack the mortar fully into the joints.

Most repointing joints are made flush, then slightly depressed with the slicker tool, then brushed. However, if the original mortar joint finish was a grapevine or concave, strike with the proper tool for this finish after the repointing is done and before the mortar gets too hard.

When large or unusually deep mortar joints need to be repointed, the best procedure to follow is to

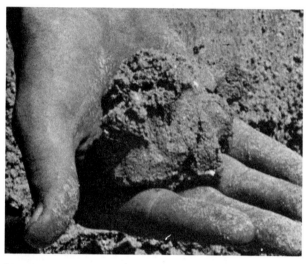

Mortar for repointing should be prehydrated, which means mixing it with just enough water to blend the ingredients but not enough to make the mix workable, and then allowed to set for 30 to 45 minutes before adding more water and applying it.

Before applying the new mortar, dampen the prepared joints with a brush and water.

Use the slicker tool to fill the joints with new mortar. Point the head joints first, then the bed joints, so that you can make unbroken horizontal strokes with the slicker.

fill in about half of the joint depth, wait until this is thumbprint-hard, and then repoint the rest of it. This decreases the possibility that the joint will crack or sag. Flick a little water on the wall with a brush and a bucket of water as the work progresses if the bricks are drying out too fast. This may be the case in hot weather, especially if the bricks are old sand bricks.

Repointing of stonework is done basically the same way, with the final joint finish tooled to suit.

If the weather is hot or dry, dampen the finished repointed brickwork with a fine spray of water from a garden hose. This will prevent the mortar from drying out too fast and cracking.

It is a good idea to wear safety eye protection when cutting out the mortar joints, especially on a windy day. If there is a lot of repointing, I recommend you rent scaffolding rather than working off of a ladder.

Mortar will always look darker when it is wet, so expect a little lighter color after it has dried.

You may wish to experiment with a small section of wall first if you're looking for a certain mortar color, before committing a large area. I would suggest repointing an area about 2' square as a test area if there is any question.

You can use a small mason's pointing trowel for filling up larger places or around doors or windows. Repointing can be done more efficiently if you have a "hawk," which is a square piece of plywood about 8"×8" with a handle nailed on it. Factory-made aluminum hawks can be bought at most

Usually, repointed joints are merely depressed slightly with the slicker and brushed when they are dry enough not to smear.

Repointing stonework. The joints should be tooled to match those in sound parts of the wall.

A partly repointed wall, showing good smooth bed joints in the completed section and properly cleaned-out joints below, ready for repointing.

building supply stores; they have a wood handle that is threaded into the center of the bottom. Mortar is placed on the hawk and picked up from there with the repointing tool.

REPAIRING A CRACK IN A MASONRY WALL

Before attempting to repair a crack in a masonry wall, try to determine what caused the crack. Generally, cracks are caused by settlement or movement of the wall somewhere. This could have been caused by the earth settling, excessive amounts of water collecting along the wall, pressure from large tree roots too close to the masonry wall, or expansion of the masonry wall because of temperature changes, etc. Whatever the cause, the crack will become worse if let go!

A visual inspection will in many cases determine the cause. Basically there are two types of cracks that develop—settlement cracks and expansion cracks. A settlement crack as a rule will be in a straight vertical line or shear, and an expansion or contraction crack will appear as a stair-step pattern.

If the settlement crack is on a fairly new house, it is best to wait for about a year to give the crack time to stop moving. Then chisel out a V shape along the crack line, about 1″ deep, and repoint with a good average masonry mortar such as has been recommended throughout this book.

If the crack is long or extremely deep and is in the brick as well as the mortar joint, be sure to wet the area first, repointing in several applications and allowing the mortar to stiffen between each application. Match the cracked brick face by mixing a

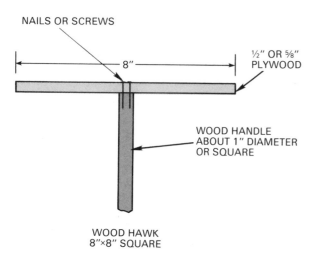

You can easily construct your own hawk, which will be more convenient to use than a trowel for holding mortar while repointing.

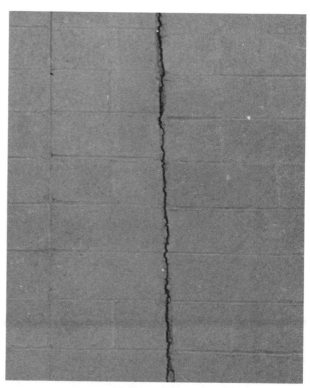

A settlement crack usually runs in a straight line from top to bottom of a wall.

An expansion or contraction crack usually runs in a stair-step pattern, diagonally.

little brick color (available from your building supplier) in the mortar, and then neatly point up the cracked area. In some cases, it may be necessary to remove portions of the wall and take out the cracked bricks, then rebuild that portion of the wall.

Stair-step cracks, caused by expansion and contraction, are fixed the same way as settlement cracks. However, you can never be sure that they will not appear again during changes in the temperature. If this does happen, then the crack will have to be caulked with a masonry caulking compound that will flex with the movement.

REMOVING OLD PAINT FROM BRICK WALLS

In restoring an older brick house to its original condition, one of the most frequent problems is how to remove the old paint from the walls. In many cases it may be an accumulation of years of paint and quite thick. Brushing with a wire brush is the simplest way to start and remove the loose scales. Sandblasting is another frequent method of removing old paint from brickwork. The danger of sandblasting, however, is that the bricks may be soft and the sand may actually cut away the surface of the bricks, making them highly vulnerable to moisture. In my opinion, sandblasting has ruined many fine old homes and should be used only as a last resort. Commercial sandblasting contractors provide this service if you care to go this route.

There is an efficient method of removing paint without damaging the brickwork. Exercise extreme care, however, in applying and removing the chemicals used, and wear safety eye protection at all times.

Following is a procedure I have used with a great deal of success:

- Obtain two 8-quart pails (plastic or rubber) and put 1 gallon of clean water in each pail.
- Wearing a pair of rubber gloves, pour 1 quart of caustic soda (lye) into one of the buckets and stir.
- Add 8 oz. regular cornstarch to the other bucket of water. Stir and blend the cornstarch and water together.
- Next, mix the water containing the cornstarch and the water containing the caustic soda together, slowly stirring all the time.
- Apply the mix on the painted brickwork from the top to the bottom of the wall with a plastic-fiber brush. If any of the solution does get on your exposed skin, flush it off immediately with running water.
- Leave the solution on the wall for about an hour. It will stick like a gelatin and cook or eat the paint

off the wall. If you encounter stubborn spots that do not come off, repeat the procedure.
- Flush the wall off with water from a hose with a nozzle attached. You can rent a high-pressure washer that will do a better job from most rental supply stores.
- Be careful that all shrubbery, plants, or objects that may be damaged are protected with plastic.

Caustic soda (lye) is available from most hardware stores, and cornstarch is available from most regular grocery stores. The paint will cause a mess when it is flushed off the wall, so be sure to rinse the area down well to dilute it and clean the residue from the ground or paving. This is the best method I have found for removing old paint from brickwork, especially oil-base paint.

REMOVING VARIOUS TYPES OF STAINS ON MASONRY WORK

If you know what kind of stain you have, you can usually remove it quite easily with the proper materials. The most common stains are from efflorescence, mortar, and smoke.

Efflorescence

Efflorescence is the deposit of crystallized salts on the face of a masonry wall. As a rule this is a whitish stain that is very objectionable on a completed wall. It is caused by salts that originate in the brick itself. The presence of moisture causes the salt to leach to the surface of the masonry as the bricks or blocks dry out. The ultimate cure is to dry the wall out completely, but this takes a long time. The best prevention is to keep moisture out of the wall as the project is built by covering the work at the end of the working period, and also by storing the masonry materials off the ground on boards or pallets and keeping them dry with a waterproof material such as plastic or polyurethane.

The best method of removing efflorescence is to scrub the area with a solution of 1 part muriatic acid to 10 parts water, being sure to wet the wall first and rinse well at the conclusion of the cleaning process.

Be sure to wear safety eye protection and rubber gloves during the washing process. If any of the solution does get on the skin, rinse it off immediately with plenty of running water, which will neutralize the chemical cleaner.

It may become necessary to repeat the cleaning process several times to remove all of the stain. Efflorescence will still not completely disappear until the wall has entirely dried out from within. Muriatic acid is available from most building supply dealers or hardware stores. Be sure to read the label before using and observe all precautions.

Mortar Stains

The same solution mentioned to remove efflorescence will clean mortar stains from bricks or blocks. Mix the same solution of 1 part muriatic acid to 10 parts water and rinse well before and after cleaning.

An optional product that is very popular with professional masons is called Sure Klean 600 Detergent Masonry Cleaning Compound. It is manufactured by Prosco, Inc., 1040 Parallel Parkway, Kansas City, KS 66104 and is available from most building supply dealers.

Prosco also manufactures a number of masonry cleaners for other specific purposes under the Sure Klean line. One such product, called Defacer Eraser, is excellent for removing general stains, including writing or graffiti, from brick, concrete block, or stone. A coat of the solution is applied to the surface, using a synthetic-fiber brush, and allowed to stand for about 15 to 30 minutes and then flushed off with water. Follow the directions and precautions printed on the label of the bottle for best results.

Smoke and Dirt from Fireplaces

Cleaning smoke and dirt from fireplaces can be done successfully with a Sure Klean product called Fireplace Cleaner. You spray it on and wipe it off. Repeat the procedure if all of the smoke stain does not come off on the first application.

The Brick Institute of America recommends a poultice formula for removing stubborn stains on fireplace fronts. It is easy to use and works well. First, mix 1 cup powdered pumice and 1 cup household ammonia and ½ cup water. Stir to a thin creamy paste. Apply the mixture to the smoked area, using an old paintbrush. Allow to dry, then scrub off with a wet scrub brush. Repeat the entire procedure if necessary. In place of the pumice, inert materials such as talc, whiting, fuller's earth, bentonite, or other clay may be used.

There are many other types of stains on masonry work that can be removed by using different chemical cleaners. The Brick Institute of America has done a lot of research in this area and for a nominal cost will send a complete list of the causes and cleaners on request. The address is Brick Institute of America, 1750 Old Meadow Road, McLean, VA 22102. Ask for Technical Note Number 20 Revised.

Remember, all cleaning products for masonry work are somewhat dangerous, and strict adherence to directions on the labels is advised.

WHAT YOU CAN DO ABOUT THAT WET BASEMENT

Dampness in a basement is a tough problem to deal with, as it is caused either by water penetration through the wall or by condensation on the inside. If the problem is condensation, the installation of a dehumidifier is the easiest cure. However, the tem-

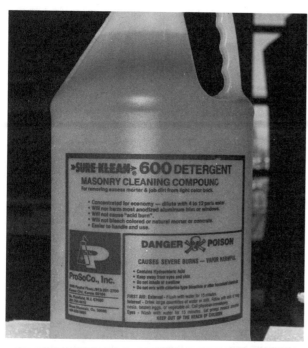

Sure Klean 600 Detergent is a convenient substitute for a muriatic acid solution when removing mortar stains.

Sure Klean Defacer Eraser is a general-purpose masonry cleaner especially good for removing graffiti and other paint stains.

perature inside the basement must be maintained about 60° or better or the coils will frost up and the dehumidifier will not work.

If the problem is caused by water pressure from the outside, you should try to determine where the pressure is coming from before attempting to remedy it. Look to see if any of the downspouts from the rain gutters are detached and running water around the wall. Make sure that a concrete splash block is under the bottom of the downspout to carry the water away from the wall area. You can also dig a trench and pipe the water from the downspout away from the house. This is very effective. Leaking soffits at the top of the house may also be the culprit, causing the water to run down the wall and into the ground next to the house. This as a rule is easy to spot, as the wall will show some stains. Whatever the problem, if water is standing around the exterior of the foundation wall, it should be corrected before you try to work on the inside of the foundation or basement.

Damp-proofing the interior of the basement wall is usually ineffective unless the problem is minor. There are on the market a number of good cement paints that help and can add color to the inside of your basement. I like a product called Thoroseal. It is manufactured by the Thoro System Products Co., 7800 N.W. 38th Street, Miami, FL 33166. It is carried by many building suppliers around the nation and is easy to use.

Specific instructions are printed on the containers of the product, and they should be followed carefully. Basically, however, make sure the wall is clean and free from loose paint or dirt before applying the cement paint. Repair any small cracks or holes with mortar before starting to paint. Let the patched areas cure for several days to ensure good results. Mix the waterproofer as directed on the bag or container and dampen the wall first, just before application. Brush on a coat, varying the direction of the brush to fill all voids in the blocks, and let dry overnight. Redampen the next day and apply the second coat.

An inexpensive homemade formula that I have used successfully for white cement paint is as follows: Mix 3 parts white portland cement to 5 parts finishing hydrated lime to 1 part calcium chloride. Let the mixture set for about 20 minutes and it will thicken to a creamlike consistency. Stir and apply to the wall with a brush. Follow the same procedure mentioned above, wetting the wall first. It is also a good idea to dampen the wall slightly with a garden sprayer after all of the cement paint has been applied to prevent it from drying out too quickly. Most

Sure Klean Fireplace Cleaner is designed for removing smoke and soot stains.

types of waterproof paints that I know of contain some portland cement. You can save some money by making your own.

SILICONES FOR DAMP-PROOFING THE EXTERIOR OF A MASONRY WALL

Some bricks, blocks, or stones absorb a lot of water, and it penetrates to the interior of the house. An especially soft or old used brick will do this. The penetration can be prevented by applying silicone to the exterior of the wall.

Masonry silicones are sold as a clear liquid that is available from your building supplier. There are two major types, water-based and solvent-based, either of which can be applied to the wall by brushing or spraying. Silicones do not cause any color change, but when applying them to a wall that has colored mortar, do a small area first to determine if any bleaching will occur. The silicone penetrates bricks to a depth of about ⅛" to ¼". It does not actually seal the openings but slows down water absorption by changing the contact angles between water and the walls of capillary pores in the masonry. This creates a negative capillary action that repels water rather than absorbing it and lets the wall breathe.

Since there are different formulas of silicone, ask your building supply dealer which is the best for your individual situation. A 5% silicone resin is considered to be average.

As was the case when applying cement paints, be sure to fix any cracks first and remove any loose particles of mortar or paint before applying. The beauty of using silicone is that since it is clear, it is invisible, yet it is highly effective. It will last five to ten years before another application is needed. The manufacturers recommend spraying rather than brushing for best results, but it can be done either way. Protect any flowers, shrubbery, etc., and protect yourself by wearing old clothing, eye protection, and a dust mask. There are many brands on the market, but all should meet federal specifications.

CAULKING

There are a number of caulking products on the market for filling up cracks in masonry, wood, or concrete. A very good one that I recommend is the General Electric Company's Silicone II. This sealant yields a durable, water-resistant seal on such surfaces as brick, concrete block, granite, limestone, marble, sandstone, and slate. It is long-lasting and flexible, and the manufacturer claims it will not crack, crumble, or dry out. It is easy to apply even in cold temperatures and can be smoothed up to six minutes after application. It comes in light gray to match common concrete and masonry surfaces and in a tube to fit your caulking gun.

If you have problems locating it, write to General Electric Company, Silicone Products Division, Waterford, NY 12188.

Silicone for masonry walls is called a repellent because it does not seal the wall but instead affects its surface so that it repels moisture rather than absorbing it.

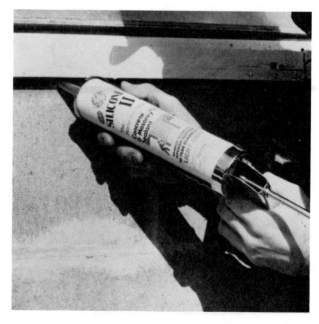

Silicone caulking compound is handy for filling cracks in masonry and between masonry and wood.

PARGING AND STUCCOING A WALL

Parging or plastering a wall is a frequent repair the handyman should know how to do. It can be on new work or on old walls.

Before starting, clean out around the bottom of the wall, using a stiff broom or brush to remove all dirt. You will need a parging or plastering trowel to do a good job. An aluminum or homemade wooden hawk to place your supply of mortar on also aids in parging, saving constant trips to pick it up from the mortarboard or wheelbarrow. Parging trowels are rectangular, like a cement-finishing trowel, but are smaller.

For parging a foundation wall, I recommend a mortar mix of 1 part portland cement to 1 part hydrated lime to 6 parts sand. This makes an excellent mortar that will stick to the wall well. Mix the mortar a little thinner than for brick so that it will be easier to trowel on the wall. Load the hawk with mortar with the plastering trowel.

Dampen the wall first with a fine spray of water from a hose. Start at the bottom of the wall and trowel a cove or angled thick coat of mortar to help drain water away from the wall.

Next, trowel the mortar up the wall with long smooth strokes, smoothing it out as much as possible to avoid unevenness. After a little practice this is not too hard.

After the parging has dried enough not to sag, scratch the surface with either a regular scratching tool or an old stiff house broom. This is done so that the second coat will bond to the first coat. Your local building supplier stocks inexpensive metal scratching tools for this purpose. They resemble a large metal comb and are called scarifiers.

Let the first coat of parging cure overnight for best results, then remoisten with a fine spray of water from a hose and parge the second coat of cement mortar on the wall. Smooth out as much as possible with the trowel, or if you prefer, you can

Before parging a wall, clean thoroughly around the base so that the mortar will bond well all the way down.

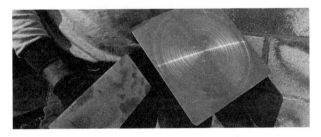

Proper tools for parging. The parging trowel, on the left, is essential for a smooth job, and the hawk again is advantageous to use.

The mortar is loaded on the hawk from the mortarboard or wheelbarrow. The mortar should be thinner than for laying brick so that it will trowel on easily.

The mortar at the bottom of the wall should be coved out as shown to direct water away from the wall.

Trowel on the mortar in long smooth strokes, starting at the cove you have made at the bottom of the wall, in as even a thickness as possible.

apply a brush finish by taking a soft-bristle brush and brushing the parging lightly.

For stucco, I like a lighter shade of mortar. The more lime in the mix, the lighter the color will be. My favorite mix for stucco is 1 part portland cement to 2 parts hydrated lime to 8 parts sand. This will stick to the wall very well because of the higher content of lime and is highly water-resistant. The stucco can be finished smooth, brushed, or slightly whipped on the surface with a lifting brushing motion. This is also a suitable mix for repairing an old stucco wall that has cracks or needs to be replaced.

There are a number of dry-packaged stucco mixes on the market—all you do is add water. They are, however, considerably more expensive if you have a large amount of stuccoing to do.

CLEANING YOUR CHIMNEY

The use of wood-burning stoves and fireplaces has increased tremendously over the last several years because of the high cost of oil and gas. Chimneys need to be cleaned after each heating season. Chimney fires result from a buildup of creosote in the flue and stovepipes. The creosote must be removed for your chimney to operate safely.

Although there are a lot of professional chimney sweeps, if you don't mind getting a little dirty and climbing to the top of your chimney, the job of cleaning is not that difficult and you can be certain that it is done right.

You will need a few tools to do the job. There is sure to be a store in your area that stocks equipment for cleaning fireplaces and chimneys. Select a brush that fits the flue lining in your chimney. This is important, because a brush that is too large may get jammed in the chimney, while a brush that is too small just will not do the job.

These brushes come with threaded lengths of fiberglass rod that screw together to fit any chimney height. The end of the brush has an eyebolt through which a rope can be tied so that the brush can be pulled through the chimney by someone at the bottom. This will remove most of the creosote and soot buildup. Although one person can clean a chimney, it is much easier with two. Your wife will make a great partner for this.

Cover any furniture in the room to protect it and you are ready to clean. If the stove is a fireplace insert, it will have to be moved out. Remove the stovepipe and take it outside. Use a wire brush to clean the stovepipe of all creosote and soot.

Scrape the inside of the stove, the firebox in the fireplace, or the chimney, whatever is the case, to remove all soot and creosote buildup.

Smooth the second coat carefully with the trowel, or give it a brush finish if you prefer.

The first coat is scratched with a so-called scarifier so that the second coat will bond well.

A loop or eyebolt at the end of a flue brush allows a rope to be dropped down the chimney so that someone at the bottom can help move the brush back and forth in the flue.

Tie a rope to the eyebolt of the chimney brush and drop it down inside the chimney flue from the roof adding sections of rod as you go. With the help of someone at the bottom, work the brush up and down the chimney, pulling it back and forth. A flashlight or a small mirror can be used to shine down the chimney to determine when it is clean.

Use an old vacuum cleaner to remove all of the soot and creosote that has collected in the base of the fireplace, on top of the smoke shelf above the fireplace damper, or at the cleanout door. If you inspect the chimney with a light and find it is still not clean repeat the cleaning process. Stubborn places in the flue where there is a buildup of creosote can usually be removed by holding a metal chain down the chimney from the top and swinging it gently around in a circle. The chain will flail against the edges of the flue and cut the creosote away. If a chimney is cleaned each heating season the creosote does not get a chance to build up and it is relatively easy to clean, although it is still a dirty job.

The front of the fireplace can be taped with newspaper during the brushing process, but if you want to use a helper to help move the brush back and forth, the front must be left open.

REPAIRING A CHIMNEY TOP

Defects in chimneys are often neglected because they are not easily seen. Periodically the top of a chimney should be inspected by either examining it with binoculars or climbing a ladder and getting on the roof.

Check for eroded mortar joints, cracks, and loose bricks. If the only problem is worn-out mortar joints, they can be repointed as described earlier in this chapter. If you don't mind working on the roof, this is a fairly simple task. If the flashing needs replacing, I strongly recommend you contact a carpenter or roofer, as it has to be installed correctly and the job is not easy.

If the chimney top is in serious condition, it should be torn down to the roof and rebuilt, with new flue linings. You can call in a professional mason or do it yourself.

If you are going to rebuild it yourself, build a scaffolding around the chimney first to work on and to protect the roof. Cut vertical scaffolding legs to fit the pitch of the roof, and brace and nail them together with other lengths of lumber or boards. Wood 2×4s or 2×6s will do the job. Cover the area with a tarpaulin to protect the roof surface. Place a scrap of plywood under the legs to protect the roof

The scary part. Think ahead to make this job as safe as possible; wear nonskid shoes and position ladders properly. That TV antenna mast might not hold if you lost your balance and grabbed it.

Clean not just the flue but all other parts of the system— stovepipe, the inside of the fireplace or stove, the shelf above the fireplace damper, and the cleanout area.

A chimney top in as bad shape as this should be torn down to the roof line and rebuilt. Check to make sure the chimney below the roof does not need rebuilding too.

before going to work. The same care should be taken to protect the rain gutters or objects below. A loop of rope tied around the chimney is a good idea to keep the scaffolding from slipping. If you set up any ladders, tie them to the building also to prevent slipping. You will, of course, need the help of another person to do the job, as it would be too tiring to run up and down the ladder for materials.

After the scaffolding is in position, tear down the chimney to where it is solid. If this is below the existing roof line, then I would call in a professional mason, as this job is more than you will want to attempt. A bag of straw attached to a rope can be pushed down into the chimney, wedged snugly, to prevent old mortar or chips of bricks from falling down inside the chimney. Tie this rope to the scaffolding.

I advise not trying to reuse the old bricks unless you are sure they are solid and good.

If possible, try not to disturb the original flashing unless it is too bad. In that case, as mentioned before, contact a roofer or carpenter to replace the flashing. Make sure that the chimney is torn down to a solid base before rebuilding it.

When you have cleared away all the bad brickwork, build the chimney back up to its original height, installing new flue lining for the most fire protection. Set the flue linings on projected pieces of bricks built into the chimney for support. The top of the flue lining should extend about 6″ higher than the masonry. If there is a television antenna attached to the chimney, I strongly recommend you not put it back but provide a separate mount. Often a small chimney will break loose because of winds whipping the antenna.

The top of the chimney should be sealed off with an angled wash coat of mortar to prevent moisture from getting into the brickwork. The mortar for this should be a little richer than used for laying the bricks, and it should be troweled to a smooth finish.

When the building process has been completed, remove the straw bag, trying not to let the loose debris go down the flue. Caulk or tar around the flashing to complete the job. If you can tackle this job yourself, you will save many hundreds of dollars.

Remember, the top of the chimney should be at least 2′ above the peak of the roof to draw properly.

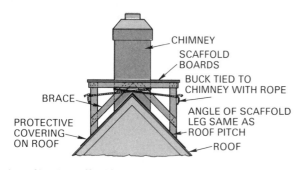

A roof buck scaffold for repairing a chimney. Note the rope holding the scaffold in place; otherwise it could slip and tip the mason off the roof.

In this case the chimney had to be torn down below the roof, and new flashing will have to be installed—a job for a professional. However, you can do the brickwork yourself.

Be sure to tear down the old chimney until you reach sound brickwork.

Once the chimney is built back up to its original height, seal off the top with a neat angled coat of rich mortar.

One last caution: Before building a fire in the fireplace or stove, clean out at the bottom to remove any soot or particles that may have dropped down the chimney during the rebuilding process or you may have a real mess.

FIXING A SUNKEN BRICK OR STONE WALK

A common problem is a brick or stone walk that has sunk from its original position. If the walk was laid in mortar on a concrete base, then the concrete has sunk also and repairing the walk will be more of a job.

The first thing is to try to determine what caused the walk to sink. A visual inspection may give you the answer. Make sure that the downspouts from your rain gutter do not empty and run under the walk, as this will certainly undermine it. If the walk is rather new, then it probably was laid on filled earth and is still settling. In this case, I would wait at least a year before doing any repair. If a contractor built your home, he should be responsible for a period of one year, so contact him as soon as the problem is noticed and the repair should be completed within that time.

If the walk has been in place for a number of years, remove the bricks and clean them off. Then pour an additional concrete base to the height desired, taking into consideration the height of the bricks when they are relaid in mortar. Lay some concrete reinforcement wire in the new concrete for extra strength. After the concrete has cured for at least 24 hours, the bricks can be relaid in fresh mortar.

If the walk was laid in gravel, stone screenings, or sand, it will be relatively easy to fix. Take up the bricks or stones that have sunk, add more stone screenings (stone dust) and/or sand, and relay the bricks, settling them into position with the end of the brick hammer handle.

REPAIRING OR REBUILDING A MASONRY RETAINING WALL

A common problem that occurs in masonry around the home is a retaining wall that no longer restrains the earth behind it but is falling over. Generally, the job won't be easy physically, as it is hard work. There is no way that you can push the wall back into position, brace it, and expect it to look right or stay there. The only proper method of repairing it is to tear it down and rebuild.

If you have such a problem with your retaining wall, tear it down and try to salvage as much as possible of the masonry materials, as you may be able to use some of them over again. Next, try to determine what caused the problem before starting the rebuilding process.

There can be a variety of causes. Maybe the wall was not built thick or wide enough and just could not take the pressure. Or maybe no drains were built through the wall to relieve pressure from water that naturally collects behind the wall. Or downspouts from your rain gutters have been emptying behind the wall and dumping a tremendous amount of water, building up pressure. Or perhaps the wall was never built on a footing or below the existing frost depth for your area, and you will have to excavate and pour a footing.

I would start out by first correcting the buildup of any excess water in the back of the wall. Dig a small trench in back of the wall and lay some drain tile on a bed of crushed stone to provide drainage. Put the stone under and over the drainpipe or tile. Flexible drainpipe about 4″ in diameter and made of black plastic is stocked by all building supply dealers. Run this pipe to some place where it will drain away from the wall. Never pipe it to a sewer or septic system that serves the house, as this is a violation of the building codes in most localities. Let it drain away on the surface of the ground somewhere that is not a problem.

SAND

SUNKEN BRICK WALK

SAND

GRAVEL OR STONE FINE SCREENINGS

REPAIRED BRICK WALK

A walk that was not laid on concrete in mortar is easy to fix—just take up the bricks and add more sand, gravel, or stone screenings.

This retaining wall cannot be straightened—it must be torn down and rebuilt.

Assuming you have taken care of the back of the wall, if a downspout is empting near the wall, attach a section of drain tile to the end of it, dig another trench, and install the drain tile or pipe so that it will carry the water away.

I have discovered over the years in rebuilding retaining walls that a number of them were built on bare ground with no footings at all. If this is the case and it is a block or block-and-brick retaining wall, pour a new footing that is at least twice as wide as the wall to be built on it, and a minimum of 8" thick.

If the original wall was falling over, then it is evident that there was a lot of pressure against it. I most strongly advocate that the wall you rebuild be reinforced with reinforcement steel of some type and concrete poured in the wall around the reinforcement. This should be done at spaced intervals, approximately 6' apart. I would use ½" steel rebar for the reinforcement. In addition, I also recommend you build a 12"-wide wall, especially when there has been a problem. It doesn't cost that much more and is a lot stronger and more stable. You don't want to go to all of that work rebuilding a wall and have the same problem again.

Use a higher-strength mortar than normal. I recommend a Type S masonry cement, or you can design your own by using 1 part portland cement to ½ part hydrated lime to 3 parts sand. Type S masonry cement is sold by masonry supply dealers. You can, of course, buy 12"-wide concrete blocks or build a 4" brick wall backed up by 8" block. If you go with the brick-and-block wall, be sure to tie the brick and block together about every 16" in height with metal joint reinforcement to form a strong wall.

Build 4" concrete or tile drain tiles into the wall periodically, to allow any water that collects on the back to escape. The back ends of the drain tiles should have nonrusting metal screen (aluminum) over them to prevent soil or mud from building up and clogging the pipes. Install these drain tiles about 1' above the finished grade line on the face of the wall and give them a little slope to the front so that water will not lie in the pipe.

In addition, I would parge (plaster) the back of the wall with mortar to waterproof it. Complete the wall to the desired height and lay in mortar the last course or cap, letting it project about 1" over the front and back of the wall. This can be of bricks, stone, or concrete slabs, as you wish.

When there is a severe amount of stress on the back of the wall, a racked-back lead of block can be

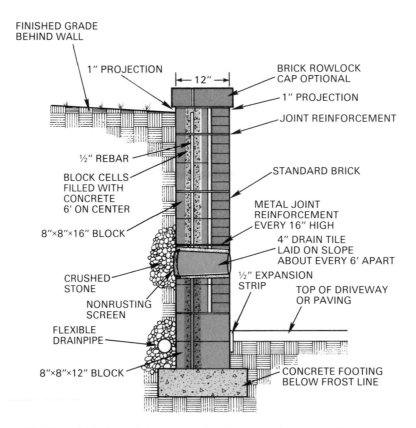

Cross section of a reinforced brick-and-block retaining wall, wherein proper drainage is just as important as strength.

built into the earthen bank to act as a brace, approximately 6' apart. It is not necessary to install a footing under these short racked-back leads, as they will be completely surrounded by earth when the wall is completed.

Dry Stone Walls

In a dry wall, no mortar or concrete is used. Many of these walls, especially on the grounds of older homes, erode and start to separate over a long period of years. Generally this was because the wall was not built to good standards in the beginning. A well-built dry stone wall will enhance the overall appearance of any property and home.

To repair or rebuild a dry stone wall, one should remember some basic rules of construction. Walls that are less than 3' high can be built right on top of the surface of the ground without any excavation at all. Just make sure that you have leveled off the area somewhat so that the stone will sit firmly. For higher walls, it makes a better job to dig a trench a foot or so in the ground and tamp it solid.

As is the case with brick and block retaining walls, excessive water may be the reason a dry stone wall did not stay in position. After checking around to determine this, you could provide some drainage along the bottom of the wall or slightly under the ground by digging a trench and laying some drain tile or flexible drainpipe on a bed of crushed stone, as for brick retaining walls. Make sure that the drain tile has a slope so the water will drain away, rather than be trapped and lie in the pipes. In most cases, drain tiles are not necessary for dry stone walls, but observe what the situation is in your particular wall location.

Dry stone walls do not have to be laid in courses as do bricks or blocks, so you have a lot more freedom to be creative and artistic. Assuming you have leveled up the area where the wall is to be built, follow these guidelines to provide a strong wall.

A dry wall should be wider at the bottom than at the top. This is known as a battered wall. Battering on both sides is more effective than battering on one side, as the wall will act as a wedge against the earth. For a wall that is 3' high, the minimum width at the bottom should be about 2'. For every 6" you build higher than 3', 4" should be added to the bottom width. I don't recommend building dry stone walls higher than 5'. The top of the wall should be one-fifth narrower than the bottom part. If the stones have good flat edges, like fieldstones, the slant can be as little as 1" to 1'. If the wall is to be built of rounded rubblestone, it may be necessary to slant the wall 3" to 1'.

For the first several courses, use the larger stones, as they not only are easier to work in place, but add more stability to the lower section. Try to lay one stone over the next or one over two, to bond the wall together. This is necessary, as no mortar is

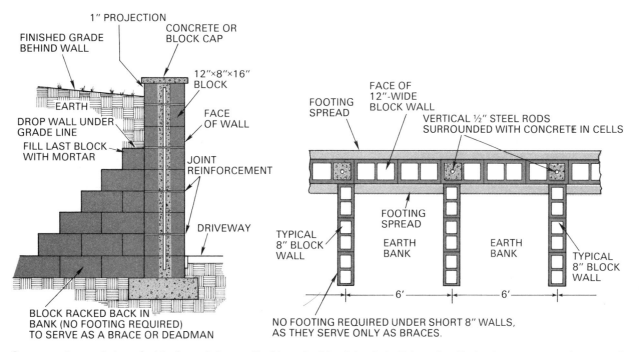

Cross section and plan of a block retaining wall with racked-back leads built into the dirt bank to serve as a brace and take some of the pressure.

being used to unite the wall. Fit smaller stones in between, until the stones will lie in the wall firmly and not shift.

As often as possible, lay some large flat stones across the wall to act as bond stones or ties. Don't worry about leveling the wall, but it is a good idea to work up the corners of the wall first, then pull a range line through to serve as a guide and keep the wall in line even though it is being built on a taper.

The stones should gradually become lighter in weight as the wall progresses. However, use a variety of shapes and lengths to make a pleasing appearance. Soil can be sifted and packed in between small stones or loose places to solidify the wall. You can also use small wedges or chips to tighten up the wall. Try not to lay one stone directly over the one beneath, and avoid vertical joints or edges in line.

This is a common rule for any stonework! The secret or trick of building a good stone dry wall is to take your time, have patience, and fit the stones tightly in position. Once you get a loose place or wobble, there is the possibility the stones may work loose later.

I have found the easiest method of building a battered stone wall is to make a wood guide board on the batter desired, and plumb the straight edge of it with a level as the corner is built. Once both corners are built to the guide board, the line is attached and the stones are laid to the line.

Select stones with the best edges for the ends, and be sure to tie the corner stones into the wall by bonding one over the other, as this will be the weakest point.

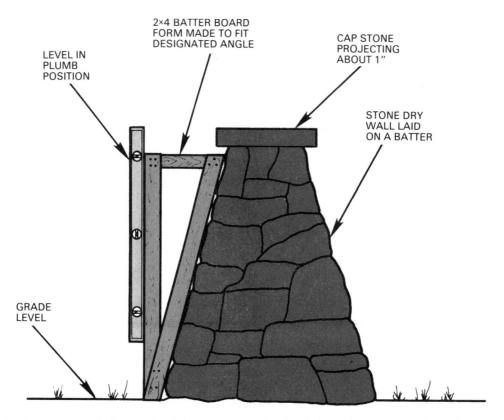

You can maintain the taper of a battered wall by using a wood guide, built to the proper angle, and a level.

INDEX